Alexander von Millern

All about Petroleum, and the Great Oil Districts of Crawford

and Venango Counties, Pa.

Alexander von Millern

All about Petroleum, and the Great Oil Districts of Crawford and Venango Counties, Pa.

ISBN/EAN: 9783744762656

Printed in Europe, USA, Canada, Australia, Japan

Cover: Foto ©berggeist007 / pixelio.de

More available books at **www.hansebooks.com**

ALL ABOUT

PETROLEUM,

AND THE

GREAT OIL DISTRICTS

OF

CRAWFORD AND VENANGO COUNTIES, PA.

⁕

THE MOST COMPLETE AND MOST RELIABLE
DESCRIPTION OF THIS REMARKABLE
REGION EVER ISSUED.

⁕

BY

ALEXANDER VON MILLERN, LL.D.,

PROFESSOR OF MINERALOGY AND GEOLOGY, CORRESPONDING MEMBER OF THE
ROYAL AND IMPERIAL ACADEMY OF ARTS AND SCIENCES AT VIENNA,
AND OF THE GEOLOGICAL SOCIETY AT FREIBOURG, &C. &C.

⁕

New York:

AMERICAN NEWS COMPANY, PUBLISHERS' AGENTS,
121 Nassau Street.

1865.

ALL ABOUT
PETROLEUM,

AND THE

GREAT OIL DISTRICTS OF PENNSYLVANIA,
OHIO, WEST VIRGINIA, ETC.

————•◆•————

ROCK—OIL, OR PETROLEUM.

The discovery of gunpowder supplanted the old
system of warfare, with all its cumbersome siege in-
struments; the discovery of printing by Guttenberg;
of the power of steam and electricity, and the discovery of the gold fields of California, exercised their influence upon the social history of the world; while
more recently, rifled cannon and iron-plated monitors
have again revolutionized the system of warfare of the
last century,—and now the discovery of a new material—petroleum—comes to exercise a yet incalculable
influence upon the course of all industrial pursuits,
exciting, at the same time, the attention of capitalists
and others, not only to the product of the rock oil regions of this country, but also to these regions themselves, which are believed to extend from the southern portion of the Ohio valley to Georgian Bay on

1

Lake Huron in Upper Canada, and from the Alleghanies in Pennsylvania, to the western limits of the bituminous coal fields in the vicinity of the Missouri river, embracing an area of about fifty thousand square miles, a vast amount of which is, of course, undeveloped.

WHAT IS PETROLEUM

Many speculations have been indulged in about the origin of petroleum. Some think that it is the work of a coralline insect which exists underneath the coal formation in the rocks. Others insist that it was distilled by the heat of the eastern slope of the Alleghany mountains from the anthracite coal, and thence flowed to the western slope of those mountains. Others again believe that petroleum is produced from layers of coal which are submitted to a low heat, and thus a gas is evolved, which being mixed with water soaking through the crevices, becomes condensed. Many impressions of fishes having been found in the rocks, some have held the idea that petroleum comes from fishes and reptiles, destroyed by some geological change in the present oil districts.

Another theory of the formation of petroleum is, that petroleum being known to be a hydrocarbon, composed of two gases, these gases are primary elements, indestructible and exhaustless in quantity. One of them—hydrogen—is constituted of water, and of course is as exhaustless as the ocean. The other is a constituent in all vegetable forms and in many of our rocks. One hundred pounds of limestone, when burned, will weigh but sixty pounds. The part driven off by burning is carbonic acid gas. Underlying the Oil Rock is a stratum of limestone of unknown thickness, but known to be upwards of one thousand feet in depth. The water falling on the surface and percolating through the porous sandstone that overlies the oil several weeks that they recurred with singular regularity a few minutes after eight o'clock in the evening when the engine was forced to stop for twenty minutes or half an hour.

PETROLEUM IN THE EAST AND WEST.

In the almost unchanged horizontal posture of the western coal measures no considerable fracturing or fissuring took place. Faults of all kinds are uncommon, and very small when they exist at all. The rise of the stratification from the Alleghany river towards Lake Erie is a fraction of one degree. The original contents of the rocks have therefore been preserved. Not so with the anthracite basins on the southeastern side of the great coal area. Crushed and upturned and overturned, contorted and fractured in every part, this part of the earth's crust has been dried and hardened, and exposed to chemical action from the superincumbent drainage waters, until its various formations—the coal beds included in the number—have been metamorphosed and partially recrystallized.— The oils which they contained have been lost by dissolution and evaporation. The bituminous coals have become anthracites, and the last oil spring on the headwaters of the Lehigh, the Schuylkill, the Juniata, the Potomac, or the New river, ceased to flow many millions of years ago. In the west, on the contrary, in equally ancient, nay, in identically the same rocks, the petroleum still remains, having had no outlet; always hermetically sealed and under pressure. It remains partly condensed in coal beds and black shales, partly distributed through the sand rocks and limestones, and partly filling up the joints which the

1*

shrinking of ages has produced. Possibly a small portion of it may be held in caverns through the more soluble limestone strata. Especially important are the water-bearing horizons.

THE GREAT RESERVOIRS OF PETROLEUM.

The vertical cleavage pianes and few down-throw fissures which exist play but a subordinate role to these. Rain-waters percolate from every hill surface and valley bed, sidewise and downwards, leeching every permeable stratum that will give up its salt and oily contents. Along the outcrops of every coal bed issue innumerable springs of painted water. At the base of every great sand rock, and on the top of the clayey deposits next below it, collect the mixed proceeds of the drainage in a standing sheet of oily brine. Capillary attraction and hydrostatic pressure perpetually re-enforce the reservoir. The weight of rock on top and the pressure of disengaged oil-gas sends its filaments forward and upward by every secret crack to the surface again, holding it in every part ready for an explosive rush into the air when an artificial outlet is provided. If there be no fissure in the locality, the oil wells descend to the sheet of water at about the same depth. Where fissures intercept them they are of various depths and fortune, for a well may pass a fissure where its walls are polished and tight together. A well may also pass the water sheet where some change in the porosity of the rocks above and below has taken place to oppose a like obstruction. In some parts of the western coal field the dip is as high as five degrees, and the basins from five to ten miles wide. Sharp flexures make local dips of thirty degrees or more, and a central subanticlinal is sure to subdivide the basin. In the secondary basins thus formed the wells are more perfectly artesian as to the salt water ; but it is upon the subdividing anticlinals that the gas and oil collect. In such regions it is asserted that *all the blowing* and many of the spouting wells are ranged along the summits of such anticlinals. In the case of some of the old gas-blowing salt wells, their actions demonstrate that they have been bored past one gas bearing stratum to another deeper salt water stratum ; for when the water is allowed to rise in the auger hole by stopping the pumps awhile, then the gas and oil no longer come up, the brine stopping their issue. In the case of neighboring wells of different depths striking a slanting fissure, the one which strikes it highest up will deliver gas ; another striking it lower down, will deliver oil ; a third, striking it still lower down, will deliver nothing but salt water.

DANGEROUS QUALITIES OF PETROLEUM.

The compressibility of coal-oil gas is one of its most dangerous qualities, increasing indefinitely the dangers of those explosions which annually cost so many valuable lives. Confined in the walls of the gangways and rooms, it issues from innumerable cells or pockets, the larger of which are called " blowers " ; sometimes with the noise of heavy rain, sometimes with small reports. It collects among the timbers of the roof, in the upper galleries of the mine, in deserted portions of the colliery, and especially in those refuse accumulations of coal and slate called " gob" or " goaf," with which the miners pillar up the superincumbent rocks. These acres of worked out and filled up galleries become vast reservoirs of fire damp. The gas collects especially over the anticlinal rolls. From these great powder magazines, solicited by the least diminution of barometric pressure in the atmosphere, the gas rushes out to fill the working rooms. Long experience has shown that a falling barometer and explosions in coal mines always go together. But the mischief is accumulative. The vacuum produced by the first explosion is a new provocation to the world of back gas to leave its hiding place, come forward afresh, and produce another, and yet another, until the proportion of air to gas becomes too small to make an explosive mixture ; so that, like the stroke of lightning, the coal mine explosion is not a unit, but a series, cause and effect reciprocally acting to produce the last result.

THE LIMESTONE FORMATION.

The petroleum which fills cavities in the Montmorenci rocks is still unhardened. It flows in drops from a fossil coral of the Birdséye limestone there ; and at Pakenham it fills the cast moulds of large orthoceratites in the Trenton limestone to such an extent that a pint has been poured out of one. It is, perhaps, from these lower silurian fossil coralline limestones that the oil makes its way to the surface through the overlying Loraine shales to form the Guilderland oil spring near Albany, according to Beck, through the Utica slate on the Great Manitoulin island, and through the red Medina shales at Albion mills, near Hamilton, according to Mr. Murray.

The next great limestone in the ascending series is the Niagara, and Eaton early made known the oozing of petroleum from its fossil casts Hall describes it in Monroe county as a granular crystalline dolomite, including small laminæ of bitumen, which give it a resinous lustre. Bitumen sometimes flows like tar from the lime-kiln. The corniferous limestone, next above the Niagara, has the cells of its fossil corals filled with petroleum, the remains of the gelatinous coral animal which inhabited them.

The oil springs of Enniskillen, as well as the lake of solid bitumen in the same township, half an acre in extent and two feet thick, no doubt have their deep-seated sources not in the black shales of the region, but in the corniferous limestone underneath. These black shales belong to the base of the Portage and Chemung group. The wells sunk in them soon strike the argillaceous shales and limestones of the Hamilton group, and go through them toward the corniferous limestone, specimens of which yielded to Hunt's analysis from 7.4 to 12.8 per cent. of bitumen, fusible and readily soluble in benzole.

OTHER FORMATIONS CONTAINING PETROLEUM.

In the blackish Marcellus shales, at the base of the Hamilton group, are found septaria or nodular concretions containing petroleum. The same phenomenon recurs at the top of the Hamilton group. Still higher up, the Portage and Chemung sandstones [formation viii,] are often bituminous to the smell, and contain petroleum in cavities, or hardened into solid seams. A calcareous sand rock in Chatauqua county contains more than 2 per cent. of bituminous matter. These are the rocks around the famous oil springs of the Seneca Indians. It is only necessary to ascend the series of these devonian sandstones to their upper part among the rocks of the Catskill group, or just beneath them, to find oneself in the oil regions of northern Pennsylvania and Ohio, described by Dr. Newberry and others, and sufficiently treated of in the foregoing pages.

There only remains to be noticed that anomalous de-

posit of the Albert coal in New Brunswick, made famous by long litigation and the discussion of geologists
described dy Professor Dawson in his Acadian Geology, and called by Dr. Wetherill, of Philadelphia, Me-
lan-asphalt.

Its position has been misinterpreted by several ob-
servers, who have reported it a volcanic injection of
bitumen into a fissure of the earth, many feet in width,
by the force of which large pieces of the wall rock
have torn off and carried forward in the mass. It
seems, however, pretty well made out, that it was originally a horizontal bed or lake of petroleum, hardened and covered up by sand and clay deposits of carbo-
niferous age, and afterwards upturned, bent over and
fractured so as to assume its present posture. It is
not properly a coal bed, therefore, but a mass of hardened coal oil, which can be, and, in fact, has been,
mined like a coal bed, and the product used wholly for
making gas.

THE OIL REGIONS AND COAL BASINS.

The connexion of the oil regions with the coal basins of western Pennsylvania and Virginia, eastern
Ohio and Kentucky, is, in good measure, a geograph-
ical deception. The Oil creek rocks, dipping southward, pass five or six hundred feet below the coal
measures. The nearest coal bed to the more northern
springs occurs on the highest hill-tops, many miles
away. The hills in the vicinity of some of the wells
are capped by the conglomerate mass of the coal measures at least a hundred feet thick. The shales and
sandstones of the valley belong to formations X, IX,
and VIII descending, called by the New York geologist the Catskill, Chemung, and Portage groups,
extending over all the southern counties of western

New York. The southern dip carries down these oil bearing rocks, and the wells must deepen in the same direction. Mr. Ridgeway reports the lowest oil bearing sand rock as capping the hills near Waterford, on Le Bœuff creek, and the same sandstones appear on Big French creek, full of plant remains.

The following wells show the dip in a well marked manner : The Phillipps well, on Oil creek, is 460 feet ; the Brawley well, at the mouth of Cherry run, 503 feet ; the Cornwall well 530 feet ; the Avery well, over 700 feet, and at Titusville he estimates the proper depth at 1,000 or 1,200 feet.

THE OVER—LAYING STRATA.

In the Mahoning coal oil region in western Pennsylvania and eastern Ohio, near the line, the three oil bearing sand rock strata are beneath the lowest coal bed. The " Continental" boring at Edenburg, in Lawrence county, penetrated in descending order, the following formations before it struck the oil : First, the superficial drift, 80 feet thick. Second, sandstones and shales, 200 feet thick, the bottom layers of which consisted of fetid black shales, from which coal gas blew off with violence. Third, the first white sandstone, 50 feet thick, arranged in three strata, a softer middle between harder upper and lower formations ; the whole mass said to be thin, going east, and holding abundance of gas in its crevices. Fourth, shales and slates, 45 feet thick, charged with oil and gas. Fifth, the second white sandstone, 75 feet thick, softer, coarser, and tougher, or more difficult to bore through than the first, and full of gas ; after passing through which they struck the great oil stratum, 448 feet from the surface. Crawford's boring, not far off, went down 580 feet through another shaley formation, and struck oil, supposed to come up through a crevice from the third white sand rock.

INFLUENCE OF THE SAND–ROCK FORMATIONS UPON THE CHARACTER OF PETROLEUM.

That there is an intimate connection between the character of these sand formations and the character of the oil which issues from them is indubitable. The rule among miners is, as stated by Mr. Clark in the "Proceedings of the American Philosophical Society," that the harder the rock may be to drill, the lighter in color, purer in quality, and smaller in quantity, will be the oil obtained therefrom ; and the softer the rock, the darker and more abundant the oil.

The chemist of the Canada survey, Mr. Hunt, insists strenuously "upon the distinction between lignitic and bituminous rocks, inasmuch as some have been disposed," he says, "to regard the former as the source of the bitumen found in nature, which they conceive to have originated from a slow distillation.— The result of a careful examination of the question has however, led us to the conclusion that the formation of the one excludes more or less completely that of the other, and that bitumen has been generated under conditions different from those which have transformed organic matters into coal and lignite ; and probably in deep water deposits, from which atmospheric oxygen was excluded."

PETROLEUM IN AN INDUSTRIAL POINT OF VIEW.

As for the illuminating power, and the value of the light of petroleum, many photometric experiments have conclusively demonstrated that petroleum-light surpasses that of a stearin, or paraffin-candle two or three times ; that it equals photogen, and surpasses rape-oil, while it is much cheaper than any other oil.

But it is not only as a direct illuminating material that petroleum is destined to play an important role, although it will enter at no very distant day largely into the fabrication of illuminating gas, giving to this branch of industry an entirely new direction ; petroleum will become also the great motive power of the future, depriving coal and wood of their former undisputed reign. It is already extensively made use of in the manufacture of candles, varnish, lubricating compositions, and many other articles, and hardly a day passes without new qualities and properties being found in this old, but to us new, material. Every day its utility in all branches of industry increases, while, also, every day adds to the quantity of supply—and considering all these facts, one can hardly avoid the conviction that petroleum is destined to inaugurate a revolution on the field of industry.

PETROLEUM AS FUEL.

As a fuel, petroleum enters into numerous French patents. The people of the Caspian sea mix it with clay ; the Norwegians with sawdust and clay. The refuse charcoal of the French furnaces is mixed with charred peat or spent tar, and tar or pitch is added, and the whole ground or coked. As an illuminating agent coal oil is fast supplanting the animal and vegetable oils. It has always been a lamp oil of India. It lights the streets of Genoa ; but its natural odor is so disgusting that its use in Europe was, for a long while after its discovery in Lombardy, interdicted. Since the refining process was discovered, the trade has spread to every city of the Old and New World, and the annual number of patents for new forms of lamp and new kinds of candle shows how completely the kerosenes and par affines are banishing the whale oils and tallows from the market.   The outlet for the coal oil wax in England and on the continent it said to be very large, not less than twenty tons per week being condensed from bog-head cannel alone.   "The superiority of the petroleum over the paraffine wax is admitted by consumers of every kind, insolubility being the proof of merit," The cold weather in America is favorable to this manufacture.

AMMONIA AND BENZINE.

Ammonia is extensively manufactured from the English gas-water, which contains it in combination with the volatile acids, sulphuretted hydrogen, and carbonic acid, and in the form of chloride ammonium. Ten gallons of this water are distilled from a ton of Newcastle coal. The sulphide and carbonate are reduced by muriatic acid. 4,000 tons of the crystallized muriate are made annually in England. Benzine or eupion, one of the products of coal oil distillation, more explosive than turpentine, has supplanted the latter in the arts since the great rebellion has diminished almost to nothing the production of the southern pine forests.   Hence explosions and conflagrations are more numerous.   The demand for benzine in England has become unlimited, especially in early and late spring. American, *i. e.*, United States petroleum, containing much more benzine than Canadian petroleum, rules higher on that account.

PECULIAR QUALITIES OF PETROLEUM.

Petroleum, when shaken, yields a few bubbles ; but they sooner subside than in almost any other liquid, and the liquor resumes its clear state again almost immediately. This seems owing to the air in this fluid being very equally distributed in all its parts, and the liquor being composed of particles very evenly and nicely arranged.

The extensibility of this oil is also amazing. A drop of it will spread over several feet of water, and in this condition it gives a great variety of colors, that is, the several parts of which this thin film is composed, act as so many prisms.

The most severe frost never congeals petroleum into ice, and paper wetted with it becomes transparent as when wetted with oil; but it does not continue so, the paper becoming opaque again in a few minutes, as the oil dries up.

The bitumen employed by the ancient Babylonians, instead of mortar, to unite the sun-dried bricks in their colossal structures, was evidently petroleum; and the state in which the mighty ruins still exist shows how imperishable a cement this material afforded.

MEDICINAL PROPERTIES OF PETROLEUM.

Petroleum of various shades, from the green of the Barbadoes springs to the pale yellow of Amiana, has been long known to possess certain medicinal properties. The rock oil of Barbadoes, or as it has been vulgarly but improperly called, Barbadoes-tar, has been found a useful stimulant to torpid bowels, promoting in such a temperament the alvine discharge. The petroleum found at Gabian, near Beziers, in France, has been called Olean Gabianum. It has been given as an excitant expectorant; and mixed with tincture of assafœtida, in tapeworm. Lucas, a German physician, recommended it both inwardly in the form of emulsion, and externally in the way of friction over the abdomen, as an effectual means of curing tape-worm.

The chief value of petroleum, however, is as an external remedy in a variety of cutaneous affections. But petroleum, either by itself, or combined with any of its solvent essential oils or spirit, would in general act rather as an irritant and rubefœcient upon the skin in such cases, than as a purifying, cleansing, and soothing application.

In this dilemma the idea occurred of incorporating the green rock oil with fine curd soap. Thus a truly balsamic compound has been obtained. When the soap, used with water in the usual way, has cleaned out the cutaneous pores, a film of the petroleum is deposited in them, powerfully remedial in many of the morbid affections of the skin

PETROLEUM AS A TOILET ARTICLE.

Such petrolized soap has been found to be quite a specific in the prickly heat of tropical regions, and of equal efficacy in the fiery eruptions incident to many persons in temperate climates. Hitherto, no method had been devised for modifying efficaciously the alkalinity of soap, which being, in the best white curd article, a definite saline compound of stearic acid, and soda in its most caustic condition, to the extent of six per cent, cannot fail to excoriate delicate skins. By the happy invention of compounding petroleum with soap, each particle of that salt is enveloped with a film of balsam, which mitigates its irritant without interfering with its detergent quality. Hence we may account for the preference given to the petroline soap by all who habitually use it at the toilet-table.

SALUBRIOUS EFFECTS OF PETROLEUM VAPOR.

The number of persons handling coal oil is estimated at from 30,000 to 50,000. Its effect upon the health has been a subject of much speculation. Mr. E. G. Kelley, a chemist, says that his men sleep and live in the factory, and enjoy remarkably good health, some of them becoming fleshy and robust who were not so before. Weak lungs and asthmatic constitutions find great relief from inhaling the petroleum vapor.

MANUFACTURE OF GAS IN THE FORM OF COAL OIL.

A well-known gas engineer in London proposes that the manufacture of gas be carried on in the immediate vicinity of the mines. Here the coal is to be submitted to distillation in the simplest manner, and the products collected in the form of coal oils ; the oil so obtained may then be submitted to purification from the nitrogenous and sulphur compounds which are so fruitful a source of complaint, when they find their way into illuminating gas ; it being thought far easier and cheaper to remove all the nitrogen and sulphur from a gallon of coal oil, than from the one hundred and fifty or two hundred feet of gas of which it is the representative. When the oil has been properly prepared, and purified from all deleterious substances, it is to be conveyed to the place where it is needed, and there converted into gas. The work necessary for this purpose need only consist of a few retorts and a gas holder or two. The retort being heated to redness, a little of the oil is allowed to flow into it ; when instantly it is converted into permanent gas, and carried through a pipe into the gas holder, of the ordinary construction, from which the illuminating gas is supplied by the mains, as heretofore.

STEAM FOR OCEAN STEAMERS— NEW USES FOR PETROLEUM.

The subject of the expense of fuel for the use of ocean steamers has been an all-important one whenever a line of trans-Atlantic steamers has been proposed. We learn that an ingenious mechanic of Meadville, Pennsylvania, is engaged in experimenting upon a plan to produce from naphtha, or the residuum of petroleum, an article of fuel that will be used, at an immensely reduced cost from coal, for generating steam on board steamships traversing the ocean. The experiment is being practically tested at the Downer refinery in Corry, Pennsylvania ; where it was giving much satisfaction, producing a heat as powerful and regular as any ever produced from either bituminous or anthracite coal. It must be remembered that this article is produced from what was at first rejected as the debris or useless residuum of petroleum, but is now coming into market as one of the most valuable products.

NEW COLORS FROM THE RESIDUUM.

Among the most favorite colors for silk goods, ribbons, &c., in the market, is a color produced from the residuum of the petroleum and manufactured at the Humboldt refinery, near Plummer, in the Oil creek region. It is a bright and fixed cerulean blue, or perhaps a shade darker, but still as brilliant, and is called the Humboldt color. The process of manufacturing it is kept a profound secret by the discoverers, who are German chemists, and do not speak, if they understand English. No stranger is allowed to enter their works, except by special permission. It is stated that the Humboldt Company produces these colors from a combination of naptha and tar. The refinery uses twenty-six stills, and probably three to four hundred barrels of petroleum per day when running at full capacity. Another delicate and fashionable color, a light blue, called "Azuriena," is produced from petroleum, as well as the now famous and popular color "Magenta" and we hear that still another color called the "Rosina," is in course of production from petroleum. These colors are ascertained by dropping the oil in a certain state into water, by which the most beautiful hues are brought out. It is confidently believed that petroleum will in some shape form one of the most valuable ingredients combining the most charming tints ever transferred to canvas. Then there are many articles known in commerce as benzine, benzoin, naphtha (which is used instead of turpentine), the lubricating oils, paint oils and we do not know how many other productions, all having a base in this wonderful treasure of the earth. The residuum has recently been purchased in large quantities by parties in Williamsburg, for some purpose not known to the public. It is also regarded with favor as a disinfectant agent, and it is said that no case of yellow fever, smallpox or other epidemic has been known to exist, where it is produced in large quantities. In Italy, upon a recent occasion, petroleum was used with entire success in driving away an epidemic that was raging in a town. The objection to petroleum works being located in cities is being removed. The Jersey City Council, in which city there are a number of petroleum works, or refineries, has withdrawn its opposition to their being located within the city limits.

TRANSPORTATION AND STORAGE OF PETROLEUM.

At first, it was thought that petroleum was highly combustible, and many Governments issued restrictive

orders regarding its transportation on ships and railroads, and its storage in buildings. The English Par-
liament went even so far as to declare that petroleum
should be stored at a distance of at least seventy-five
feet from all neighboring houses. Other governments
and municipalities, principally in sea-ports, followed
this example and issued more or less stringent orders,
but most of them have been rescinded, for it has been
proved that only crude petroleum is inflammable from
the extremely volatile substance, the so-called naptha,
which is mixed with it; while the rectified oil only
equals in combustibility alcohol and turpentine, and
is even less dangerous than ether and sulphuric acid.
The city authorities of Liverpool instituted trials, and
it was found that ignited petroleum may be extin-
guished with water and some of the patent fire annihi-
lators. However, where large quantities of refined
petroleum are stored, a careful ventilation is necessary, in order to get rid of the gases evaporating from
the oil.

THE HISTORY OF COAL OIL AND ITS DISCOVLRY.

When we speak of the discovery of coal oil, in reference to late events, it must not be mistaken for
a modern invention. The extraordinary attention
drawn upon it by the discovery of a more abundant
supply, by artificial wells, since the August of 1859,
has made its previous history of comparatively little
interest to one class of minds, but, on the other hand,
has invested that previous history to philosophic eyes
with all the charm of an archælogical investigation.
Did not the builders of Babel use clay for bricks and
slime for mortar? It is evident from an examination
of any of the ruins of Mesopotamia, that asphaltic mortar was the bed into which their alabaster wainscot

pieces were set, and with which their vast terraces
were compacted, and probably their roofs protected—
the use of which so abundantly, only facilitated their
destruction when the torch was applied. The pitch
used was made by evaporating petroleum. That of
Babylon we know was obtained from the sulphur,
brine, and oil springs of Is, the products of which are
still sold in the village of Hits. The story of the cat-
astrophe of Sodom and Gomorrah, if not originated,
was perpetuated by the vast accumulations of rock oil
in the centre of the Dead Sea, as on the surface of a
heated, simmering brine vat, where it is hardened by
oxydation and drifted to the surrounding shores. A
similar phenomenon—a lake of pure petroleum elicit-
ed the amazement of the Spaniards who discovered
Trinidad.

THE GREEK FIRE A COMPOUND OF PETROLEUM.

Oil springs, in fact, have been known and esteemed,
and even worshipped, in every age and many coun-
tries. Herodotus describes a bitumen spring in Za-
cynthus, Zante, one of the Ionian Islands ; and probably this spring sufficed the Egyptian nation for their
incessant religious use of petroleum for mummies, the
embalmment of which is amusingly described in Hunt's
Merchants' Magazine for 1862. The "Greek fire" of
more modern times was probably compounded of petroleum from the Zantean springs. Dioscyrides tells
us that rock oil was collected in Sicily and burned in
the lamps of Agrigentum. The classic home of naphtha is Baku, a high peninsula on the western shore of
the Caspian Sea, containing thirty-five villages and
twenty thousand souls, rocky and sterile, without an
attractive spot, without a stream, without one drop of

sweet water except what falls directly from the clouds, and without a tree. But coal gas rises everywhere from a soil saturated with naphtha, and numerous volcanoes in action discharge volumes of mud. From the time of Zoroaster the naphtha of Baku has been sent all over Asia for the service of the sacred fire of the Parsees. The liquid streams spontaneously through the surface, and rises wherever a hole is bored. But especially at Balegan, six miles from the capital village, the sides of the mountain stream with black oils, which collects in reservoirs constructed in an unknown ancient time, while not far off a spring of white oil gushes from the foot.

A LAND OF FIRE.

Upon their festival occasions the people pour tuns of this oil over the surface of the water in a bay of the Caspian, and then set, as it were, earth, sea and sky in a blaze of light. Sometimes far grander exhibitions take place naturally. In 1817 a column of flame, six hundred yards in diameter, broke out near Balegan, and roared with boiling brine and ejaculated rocks for eighteen days together, until it raised a mound nine hundred feet in height. Of course, the population use the oil for light and fuel, and coat their roofs with it. A clay pipe or hollow reed steeped in lime water, set upright in the floor of a dwelling, serves as a natural and sufficient gas pipe. The Ghebers bottle it for foreign use : the Atecshjahns fire with it their lime kilns and burn their dead. No wonder the religious sentiment of oriental mystics was entranced by such a land of fire as Baku, where in the fissures of the white and sulphurous soil the naphtha vapors flicker into flame ; where a boiling lake is covered with a flame devoid of sensible heat ; where after the warm showers of autumn the surrounding country seems on fire ;

flames in enormous volumes rolling along the mountains with incredible velocity, or standing still expectant ; where the October and November moons light up with an azure tint the entire west, and the Soghdaku, Mount Paradise, the eastern buttress of the Caucasus, covers its upper half with a glowing robe ; while if the night be moonless, innumerable jets of flame, isolated or in crowds, cover all the plains, leaving the mountains in obscurity. The Gheber and the chemist here may worship side by side. All the phenomena of distillation and combustion, under varying barometic and thermometic conditions of the atmosphere, may be studied ; for none of this general fire burns unless when captured and applied to human uses in the lamp or stove or kiln. In the midst of this devouring element—through this world in flames—men live and love unharmed, tend sheep, plant onions, sleep, are born and die, as in more prosaic regions. The reeds and grass are in no wise affected by the flowing oil or by the burning gas. In fact, Rottiers, the traveler, thought the whole phenomenon electric, when he noticed that the vacuum in his thermemeter tube seemed to be especially full of flame, and that the east wind put to quiet the whole exhibition ; with which fact we may compare the curious discoveries of Moffat with his phosphorus thermometer, published in Silliman's Journal, Dscember, 1862, p. 437, as bearing on his theory of two normal opposite air currents. From an equally remote era the Burman empire and northern Hindostan have received annual supplies of rock oil from the wells of the Himalayan valley of the Irrawaddy, through Rangoon ; and it has always been a favorite drug in the Indian pharmacopia.

COAL OIL IN EUROPE.

In Italy, the oil wells of Parma and Modena date back nearly two centuries, the year 1640 being that assigned to their discovery. The springs of Ammiano have long lighted the streets of Genoa.

In France, oil springs have been known from time immemorial at Clermont and Gabian ; and in Canton Neufchatel ; and in Bavaria, Germany.

In the English coal mines, of course, the coal-oil gas—the dreadful firedamp—was always a well known demon to the mining population ; but in 1659 Shirley, perhaps first, describes it to the reading public as an illuminating gas. In 1733 Sir James Lowther laid pipes along the mines and burned the gases at the surface of the earth. But the lighting of London streets and houses with gas came not till 1842. Twenty years have elapsed, and there are in Great Britain and Ireland 1015 gasworks, with a capital of ninety millions of dollars, charging an average of one dollar and eighty cents per thousand cubic feet to small consumers, and deducting from five to thirty per cent. for heavy consumption. Some of these companies pay twelve twelve per cent. dividends, and many of them ten per cent. The average capital of British gas works is said to be nearly twenty per cent. less than that of American companies.

PETROLEUM IN AMERICA.

In America the history of coal oil commences with the use the white settlers found the Indians made of it for medicine, for paint, and for certain religious ceremonies. The settlers adopted its medicinal use alone and retained for more than one affluent of the Alleghany river the Indian name of Oil creek. The oil was minerals, required a manufacture, previous to the discovery of the well oil, which consisted raidically in a coking process, during which the volatile ingredients distilled over. In 1832 Blum and Moneuse patented, and Lawrent and Selligue began to investigate, the coal oil gas manufacture.

James Young, of Glasgow, in 1847, introduced the distillation process into England, and applied it to boghead cannel from which he obtained, in 1854, an annual yield of oil equal to 8,000 gallons, selling for £100,000, most of which was profit. In 1854 the Kerosene Oil Works, on Long Island, introduced the distilling process into America, and in 1856 the Breckenridge Oil Works, at Cloverport, Kentucky, distiled from cannel coal found there. By the close of 1860 there were 25 factories in Ohio; 6 in Kentucky; 8 or 10 in Virginia; 10 in Pennsylvania; 1 in St. Louis; 5 near New York city; 1 at Hartford; 4 near Boston; 1 at Bedford, and one in Portland, each averaging perhaps, 300 gallons of light oils per day; the boghead cannel yielding 75 refined from 130 gallons of crude oil per ton, and the American cannels 60 from 117. The Albert coal yields, however 75 from 110. Now that the natural crude oil issues from the earth in such abundance, the first distilling process is abandoned, and these factories are occupied in refining only.

REFINING PETROLEUM,

The refining process requires stills holding, say, 1,500 gallons of crude oil, made of boiler plate, with cast-iron bottoms two inches thick, on which the coke crust is deposited eight or ten inches deep, and is used as fuel after being removed. The heat continues twenty-four hours, and rises gradually to 800° Fahrenheit. A steady flow of oil proceeds from the end of the worm, the condensation of paraffine in which, towards the end of the process, is carefully prevented. Of the whole oil 88 or 90 per cent. comes over, and is then further purified with 5 to 6 per cent. of sulphuric acid in "agitators." Three thousand gallons of oil are kept stirred for a while, and then left to settle, the salts being tapped at the bottom. Agitated again with water, and again tapped from below, the oil is agitated a third time with strong alkali, washed again, and then transferred to the second set of stills like the first. Thus is produced, first the limid merchantable illuminating oils below 0.820, constituting from 30 to even 90 per cent. of the whole ; then follow heavier oils sold to the machine shops and railroads, or re-distilled into light oils and paraffine ; finally comes over the dark-colored paraffine oils, which, when left to stand cold in vats exposed to air, deposit paraffine in silvery scales to be itself pressed and purified with acid, hot water, and alkali. The illuminating oils are deprived of odor by standing for some days in shallow vats over alkaline solutions. Light destroys also the color, but yellowish oil at 60 cents is worth more for lamp use than colorless oil at 75.

ARE THE OIL WELLS GIVING OUT?

Whether the natural supply of rock oil will be diminished in coming time is a question of moment to the speculator, and of interest to the economist and geologist. The force with which new borings often permit the deep-set reservoirs of oil and gas to evacuate themselves, would seem itself to state the physical impossibility of its continuance ; and experience has shown that all the older wells slowly diminished their supply. Hall states, in describing the old Freedom spring, in Cattaraugus county, New York, that a well was dug fourteen feet deep, eighteen feet distant which afforded at first a large supply of oil, but soon the oil and the new springs died away together. Few perhaps none, of the old salt wells of the Sandy, the Kanawha, the Monongahela, Conemaugh, Alleghany, Beaver, and Muskingum valleys have been retained in full working condition, except by being deepened from time of time. The boring being carried further down every few years, new supplies of brime and oil and gas have been the consequence. The fiercest blowing and spouting wells of the last two years have become comparatively quiet. There is every geological reason for believing that the number and age of neighboring wells are the two elements of the calculation to determine their capacity.

When a comparison with other regions of the world is instituted, the same conclusion is arrived at. The five hundred and twenty springs of the Yananghoung, en the Irrawaddy, yield now only one hundred and twenty thousand gallons of petroleum per annum. Cases of sudden exhaustion also have occurred, when wells, beginning to blew off gas, have, in a few days, become quite dead in all respects. It is also asserted that, in every case of conflagration, the burning well has ceased its yield of oil, as if internally injured, by the cracking of the walls or by the loss of gass. On the other hand, old wells, exhausted by long practice and abandoned, have become refreshed by rest and profitable.

FALLACIOUS THEORIES.

Although the existence of earth-oil was known prior to Drake's operations, this knowledge of its existence did not amount to anything, for it was believed by peoplr generally, that it did not exist in any very large quantity ; and when the excitement did increase, and hundreds of adventurers commenced the work of prospecting, one great objection was urged against the business, and that was, that rock oil did not exist but in very limited quantities. Not only was this urged by the masses, but geologists ventured to support the same fallacious theory, and had not the pioneers been governed by a self-reliant judgment, they might have abandoned the work. The result of the determination to know by actual demonstration, proved conclusively that it did not exist in large quantities, and instead of heing a failure or a mere dream, the contrary was the case, and a temporary reverse was inflicted upon the oil fortune seekers by an overwhelming quantity of it Some were erroneously affected by this plethoric yield of grease, and as unwisely as those who did not go in, because of a belief in its non-existence, gave up the speculation because there was too much.

OIL WELLS ON FIRE.

More than once, in spite of all precaution, a spouting well has taken fire, and roared and burned like a volcano. Then pump works, engine houses, stores and boats, the soil, the stream, and the river into which it pours its flame, spread their common conflagration over day and night. In the autumn of 1861 a well about three miles up Oil creek was lit by a cigar, while thirty or forty people were standing around it, of whom fifteen were killed instantly by the explosion and thirteen severely injured. A column of fire, with its head rising and falling from thirty to fifty feet, continued to burn.

The Little & Merrick well was one hundred and fifty feet deep at first, but in the spring of 1861 was deepen-

or briefly to describe, will be confined to a section of country known as the " oil belts " of Ohio and Western Virginia, which are at the present time so greatly attractive to capitalists, and so much engaging the attention of the scientific, and those devoted to exploring the hidden treasures of nature, as well as business men and busy speculators also, of that country and of distant cities and States ; our city of " Brotherly Love" contributing her quota to make up the number. Within the last fortnight no less than forty Philadelphians have registered their names as among the " distinguished arrivals " at the two hotels—Swann's and Spencer's —at Parksburgh on the Ohio river, in West Virginia. This town and Marietta, in the State of Ohio, are the principal places of rendezvous for the so called " oil men "—-dealers in and purchasers of properties supposed to contain the precious greasy liquid. In addition to the arrivals hailing from Philadelphia, there have been registered in the same period of time gentlemen from Boston, New York, Baltimore, Washington City, Pittsburgh, Cincinnati and Oil City, also from Maine and California. The hotels are nightly thronged and overflowing with guests. Your humble servant, with three gentlemen from Philadelphia, on our arrival, one night last week, at Parkersburgh, could not be accommodated with beds. The hotels are full and every cot occupied—the steamboats on the wharf were in the same fix, berths, all of them engaged, and hence we had to resort to the parlor floor of the hotel as the next best chance. Stretched out full length on the carpet, with my overcoat for a pillow, I thus, like many ethers in the same building " with oil on the brain," I laid me down to rest, trusting that the coming morrow would bring with it some " ile strike" lucky venture by which I could make a fortune. Before the dawn of the morning a Congressman and a Squire, both fevered with the oil excitement with lamp in hand, well trimmed with oil brightly burning, sought for and found me stowed away in one corner of the room, and aroused me from a sound sleep, each requesting that I should accompany him, in different directions, to examine their oil territories. I felt desirous to serve them both, but being not omnipresent, I could not be in two places at the same time, I therefore proposed that I should take breakfast while they compromised the difficulty. But I digress. Seriously, I entertain the belief, with numbers of others who have visited and examined that part of the country, that Western Virginia is rapidly improving— that there has always been heretofore a lack of enterprise, a want of enterprising men to develop and foster the great natural resources within her boundary, to make them become useful and profitable to the citizens of that Commonwealth, now, as it were, comparatively worthless to any one. This is true, lamentably true ; but methinks apathy is dying out, the country is being aroused from its long slumbers, and that beneficial changes are constantly occuring now, especially along and contiguous to the Ohio frontier of the State—slowly but surely creeping up to her Alleghany mountain boundary, and that, before long, all parts within her limits will reap the advantages which follow a change from indolence to industry, from carelessness and shiftlessness to enterprise and pros-. perity, stepping-stones to happiness and wealth. Her geological position—her mineral contents—her soil rich in agricultural advantages, and covered with an immense growth of huge and valuable timber—her navigable waters and healthful climate, are remarkable, and combine everything that is desirable to make her a noble State.

Eastern capitalists have recently made extensive purchases in the oil territories in West Virginia, some of them amounting to hundreds of thousands of dollars. The principal oil wells in "Horse Neck" and "Bull Creek," in "Pleasants" county, are now owned in Philadelphia and New York. Philadelphians own considerable territory near Petroleum or Goose Creek and tributaries in "Ritchie" and also on Hughe's River. The Rathbone Estate, at Burning Springs in "Wirt" has been purchased by New Yorkers, and the "Eternal Centre Well" at the same place—otherwise called the "hub of the universe," which once flowed over with oil—with other territory on the tributaries of the Little Kanawha has changed hands and is now the property of Philadelphians. Many undeveloped properties in that section of country have lately been bought. I trust all of them will be attended with satisfactory results to the purchasers and to those who may hereafter become interested therein. In my next I shall describe the geological characteristics of the oil belts and some information in regard to the depths and productions of the oil wells.

PHILADELPHIA, October, 1864.

THE OIL WELLS OF OHIO AND WEST VIRGINIA.

We extract the following from the *Marietta Regis ter :*

Oil, Oil Leases, and Oil Wells form the engrossing topic of the day ; not a train or boat arrives, without bringing several persons who are desirous of investing in this lucrative commodity. At the National House, in Marietta, the headquarters for all persons engaged in the business—it is almost impossible to hear anything discussed but *oil.* Hogs have given their last oleaginous grunt, and those who have "gone down to the sea in great ships" in quest of whales, will, upon their return, find their business "up a spout." Several of the wells are blowing. Tack's, on Campbell's Run, threw, a few days since, a perpendicular stream of oil and water a distance of eight feet, through tubing two and a quarter inches in diameter, and all the surrounding country has increased fifty per cent. in value. The Coney and Gilfillan Well produces from 75 to 100 barrels per day. The Prime Well from 10 to 12 barrels. The Needler Well, within one hundred yards of Tack's, is nearly four hundred feet deep, with excellent indications. All these bases are on Horse Neck, Wood County, West Va., eight or ten miles from Marietta. On Cow Run, in this county the Bergen Oil Company are sinking several wells. Their first, 190 feet deep, produces ten barrels daily, with a specific gravity of 41; and their third, five barrels, specific gravity 40. The Newton Well, now owned by this company, has produced 17,000 barrels, and is still producing. Many other parties are boring, and most of them have succeeded in striking the oleaginous fluid. The Dutton Well, on Duck Creek, has produced 19,000 barrels, and is still yielding finely. Many wells on Paw-Paw are giving splendid yields. Long Morse, Fifteen, Eight Mile Runs, together with Duck and Little Muskingum, on this side of the Ohio, and Bull Creek, Calf Creek, Cow Creek, and Horse Neck, on the other, all promise splendidly. The average specific gravity of all the wells is about 42. The probabilities are, that this will undoubtedly become a business, one of the best paying in the country; and without much question, oil leases and oil stocks will be placed on the "Stock Board."

The Oil and Mining Business, in this neighborhood bids fair to become very large. Quite a number of companies, of heavy capital, as well as individuals, are now operating, and using a great deal of activity and energy in developing the petroleum and mineral resources of Washington county, and the adjacent territory of West Virginia. The prospects are flattering, and men of enterprise are largely investing their capital in this region. Real estate is rapidly rising in value, and the city of Marietta is receiving wealth and population—apparently substantial—in a degree that is pleasing to all its people. Men of value, in all respects, have recently come among us, attracted by resources that have been hidden in the earth, and which they are now bringing to light.

BULL CREEK OIL WELLS OF WEST VIRGINIA.

A famous region this. Twenty-four wells have been bored in this oil district. The big well of the vicinity is the " Gilfilan Well" which has yielded from 50 to 75 barrels per day for the last five months, as reported. About 24 wells have been bored from Bull Creek to French Creek and some 60 or 70 more are in process of boring. Pleasants, Wood, Wirt and Ritchie counties are rich in Petroleum deposits and at this time are commanding the attention of explorers to an unusual degree. On Horse Neck a few weeks since a well was struck that yielded 800 barrels of oil in a day. This, of course, will not be its average, although the well may give extraordinary yields. Tack & Brother, of Philadelphia, are having very satisfactory success on Rawson's Run. Their new well yielding as above is 200 feet deep. This firm has another well 380 feet deep yielding about 35 barrels daily. The reports from Duck Creek above Marietta are also very favorable to oil operators. A glance at the map will show the relative location of the Virginia and Ohio oil districts. What this phenomenon of subterranean oil in various parts of the country will upon further investigation, amount to, no one can tell with certainty. We trust the permanency of the yield will equal the hopes that are now stimulating enterprise in business circles.   We have received a copy of the Kanawha Petroleum Company incorporated under the laws of the State of New York.   The company has 416 acres in different tracts lying not far from Petroleum in Ritchie County, and very near the railroad leading from Parkersburgh to Grafton.   The location of the lands of this company is auspicious for good results to the Stockholders, oil having been found as early as 1861 on the edge of Goose Creek, within a few feet of the railroad.   The lands are covered with valuable timber and underlaid with coal and other minerals. The oil pockets seem to lie at the depths of 150 or 200 feet.

The oil in this region is said to have sufficient body —28 degrees Beaume—to answer lubricating purposes.   Upon this point we ask for more information than we now have.   We wish this and all other companies, engaged in the development of the oil deposits of the country, the realization of their seemingly wellfounded anticipations.

A TRIP THROUGH A PART OF THE OIL REGION.

The route and mode of procedure for parties from New York to pursue are :—Take the lightning train leaving Jersey City about six P. M., having early in the day secured at the principal office, 240 Broadway, through tickets to Titusville and sleeping sections. . Tickets cost $12 25, sleeping accommodations extr Trains arrive at Salamanca at eleven A. M., next day. There change cars to the Atlantic and Great Western, arriving at Corry about two P. M.   There change cars again to the Oil Creek train for Titusville.   On arriving at this thriving and bustling, but rapidly growing little town, proceed to the McCray House, the best and most popular place of entertainment—having written to the proprietor a day before leaving New York to have beds for the night, and horses for the following day engaged, or it is probable that neither can be had. Although this town bids fair before long to be of considerable importance, yet here the traveller has his first introduction to the most execrable roads—the world might be challenged to match. In the words of an Irishman, we have a graphic description of the thoroughfares :—" Sure, ain't the streets *paved* with mud, and the sidewalks *flagged* with dirty boords, full of holes." No man who cannot ride on horseback, and handle his nag well, should think of visiting this locality. Mud, mud, everlasting mud, everywhere, varied with rocks, stumps and holes of startling proportions. Does the visitor come to see some piece of land offered to him for sale, unless he has previously arranged for a conductor he will meet with trouble and disappointment in his research. Does he come to investigate the property and proceedings of some company in which he contemplates taking shares, he will be surrounded with similar difficulties. Let all plans be arranged before leaving the city. Does he come to prospect and purchase land, he has need of all the caution he can command, and must bear in mind that nearly every eligible location is bought up, and that he lacks the essential knowledge of distinguishing eligible territory from that which is, probably, valueless. Still, there are prizes to be drawn ; and, by bold and energetic action, many will yet make large fortunes, while the majority who persevere cannot fail to do well. With regard to companies, there is no doubt, if they are founded on good fee simple estates, with responsible men at the helm, that they cannot fail to do well ; while those whose assets consist in shares of the products of wells which they do not control, with a tract of worthless land in some out of the way region thrown in a blind, should by all means be shunned.

VENNAGO COUNTY.

Venango county is in the north-western part of Pennsylvania, and has an area of eight hundred and fifty square miles. The Alleghany river flows through the middle of the county. The French creek, called by the Indians Venango creek, empties into the Alleghany river near Franklin. Venango county is also drained by Oil, Teonista, and Racoon creeks. The surface is broken ; the streams flow though narrow valleys, which are separated from the uplands by steep and rugged hills. The soil of the uplands is moderately fertile, and adapted to pasturage. Wheat, Indian corn, oats and grass are the staples. In 1850 Venango county produced 98,189 bushels of wheat ; 109,042 of corn ; 255,146 of oats ; 319,870 pounds of butter ; 14,678 of maple sugar, and 15,653 tons of hay. There were thirty-one saw-mills, nine flour and grist-mills, twelve iron furnaces, one iron forge, two woolen factories, one nail factory, three agricultural implement manufactories, and six tanneries. It contained nineteen churches, and two newspaper offices. Iron ore, coal, and limestone were the most valuable mineral product ; while the creek furnishes copious and permanent motive power. The Alleghany river is navigable for steamboats, and a branch of the State canal extends from Franklin to Meadville. The county was organized in 1800, and named from Venango creek. It had, in 1860, a population of 24,974 whites and 60 colored.

CRAWFORD COUNTY.

Crawford county is in the north-western part of Pennsylvania, bordering on Ohio, and has an area of about nine hundred and seventy-five square miles. It is intersected by French creek, and also drained by Shenango, Oil, Cassawaga, and Conneout creeks. The surface is undulating; the soil generally fertile ; a large portion of it is better adapted to grazing than to tillage. Indian corn, wheat, oats, hay, butter, and potatoes are the staples. Lumber is also exported. In 1860 this county produced 387,556 bushels of corn ; 142,414 of wheat ; 418,751 of oats ; 165,662 of potatoes, and 1,267,436 pounds of butter. There were one hundred and forty saw mills, fifteen flour and grist mills, three woolen factories, two iron foundries, two wool carding-mills, two distilleries, eleven cabinet ware manufactories, three agricultural implement manufactories, and sixteen tanneries. It contained sixty-three churches, and five newspaper offices. The county contains iron ore and lime marl. It is intersected by the Beaver and Erie canal, and by the Great Western and Atlantic Railroad. The Franklin branch of the State canal also terminates in the county. It was organized in the year 1800, and named in honor of Colonel William Crawford, who was captured and murdered by the Indians at Sandusky, Ohio, in 1782. Crawford county had, in 1860, an aggregate population of 48,755, 48,573 of which were whites.

MEADVILLE.

Meadville is the capital of Crawford county, Pennsylvania, and a flourishing place, pleasantly situated on French creek, 436 miles in a west-north-westerly direction from Harrisburg, and about 95 miles north from Pittsburg. It was the principal market for the populous and fertile county of Crawford, and had quite a large export trade with grain, lumber, etc. ; but since the discovery of petroleum in Crawford and other neighboring counties, Meadville has more than doubled its former population, and is now on the high road to an importance entirely beyond the means of calculation. Among the public buildings are a handsome court-house, a State arsenal, and an academy. The elegant building of Alleghany College, under the patronage of three conferences of the Methodist church, stands on an eminence half a mile from the centre of the town. It has ten churches—two Presbyterian, a Baptist, Methodist, Unitarian, Episcopal, German Lutheran, two Roman Catholic, and one for colored people. There is also the Unitarian Theological School, founded by Harm Jam Huidekuper a female seminary, a museum, two banks, several brokers. offices, several paper mills, an edge-tool factory, etc Four newspapers are published here, and there are eight or nine hotels—the McHenry, National, Sherwood, Rupps', the Eagle, Crawford, American, and Colt's; while a new one, on an extensive scale, is in course of construction. Meadville was incorporated in 1823. For the transportation, wharfage, etc., of the daily increasing oil business, Meadville offers better advantages than almost any other place in the oil region ; while the Atlantic and Great Western Railroad Company is constructing large and substantial buildings to meet the requirements of their immense traffic.

FRANKLIN

Franklin, the capital of Venango county, Pennsylanvia, is about twenty-seven miles from Meadville, two hundred and twelve miles from Harrisburg, and sixty-eight miles north from Pittsburg. It lays on the right branch of French creek, immediately above its entrance into the Alleghany river, and is the southern terminus of a branch canal, extending from the Alleghany river to Meadville. Small steamboats run between Franklin and Pittsburg. The town looks old, and furnishes but few signs of recent improvement. Franklin contains a court house, one or two academies, two newspaper offices, and a barrel factory, which turns out two hundred and fifty barrels per day, connected with which is a patent tag factory. It has several bridges across the river and creek. Franklin not only looks an old town, but is in fact one, having been laid out in 1795, on the site of Fort Franklin. The derricks of oil wells strike the eye at every turn, and new strikes on the Alleghany river and French creek are of almost daily occurrence. It was in this town that the third oil was struck, by a blacksmith named Evans. The oil from the Evans well commands a higher price than from any other, being, like most of the wells sunk in French creek, valuable for lubricating purposes, having more gravity. Other wells are sunk near the Evans wells—viz.: two by Halingen, and two by Mann, both of Philadelphia, Simpson & West, of Philadelphia, with several others, producing more or less lubricating oil. An interest in a well on French creek recently sold in New York for $9,000.

An incredible amount of business is transacted at the Register's office in this place, which is the county seat of Venango county, reaching one million dollars per day in transferring of leases alone. The work is four weeks behind on recording cases, although a large force of extra clerks is employed. It is a pity the town would not erect a more suitable structure for a court house, when so much and such important bus-

iness is obliged to be transacted within its walls
The present court house is an old fashioned structure,
with no fire proof vault in which to deposit the valuable records entrusted to the officials.   It is advisable,
nay, absolutely necessary, that a reform in this re-
spect should be adopted.   Parties in New York, Philadelphia, Pittsburgh, to say nothing of the resident in-
habitants of the oil region, are interested in this
improvement, and unless the advice we give be soon
respected, a conflagration in the town may destroy
many valuable documents, involving many millions of
money.

The activity among the oil operators of Franklin is
not lessened by the approach of bad roads and winter.
The demand for engines and the requisite materials
for boring, is greater than at any other period of oil
history.   The feverish excitement of the early discoverers has settled down to quiet, business like operations, without any reduction of vigor or intensity of
purpose.   There is a great deal less talk and more
work than in former years.   The consequence will be
less speculation and a greater development of property.

The wells in this borough, of which about one dozen
were producing more or less oil, but none in large
quantities, have mostly been lying idle since the low
prices of last summer.   Many of these wells are now
in operation, and at present prices yield a handsome
profit.   The Brough well, at the lower end of the borough, has re-commenced pumping, and is now yielding
about six barrels.

Operations along Sugar creek are being vigorously
prosecuted, and with fair prospects of success.   At
Valley Furnace, now the property of the New York
and Pennsylvania Oil Co., two wells are going down,
and preparations are ready for another. In the vicinity
of Cooperstown some nine wells are in different stages
of progress.   The work on French creek is also pro-

gressing with satisfactory results so far ; and the same may be said of the wells along the river above and below town. The next few months will show a development of comparatively new territory to an extent not witnessed in former years.

The increasing demand for all available oil territory bring with it a daily influx of strnagers to this town, from whence they radiate to all parts of the region.

TITUSVILLE.

Titusville was originally a small, although thriving lumbering town and post-borough of Crawford county, Pennsylvania, on Oil creek, twenty-eight miles east from Meadville. It is well supplied with water-power, and had always an active trade. It was about a mile and a half from here that oil was discovered, on Oil creek, by Colonel Drake. The excitement began in 1860—61, and people came in individually and went to putting down wells, with more or less success, in some cases the first strike being the best. Since then the business has increased, and immense fortunes have been made. Among the millionaires may be enumerated the heirs of the late Captain A. B. Funk, Jonah Watson, Orange Notch, who have retired with great fortunes ; Wm. H. Abbott, Charles Hyde—all poor men orignally, except Mr. Abbott, who came here orginally worth some $40,000. Among the half millionaires may be mentioned J. W. Sherman, J. G. Hussey (living at Cleveland, but doing business here), Dr. Levi Halderman, F. W. Ames (burgess of the borough), and many others. The Dalzell brothers, formerly of Pittsburg, have a large interest here, and are esteemed very wealthy.

The visitor is not long in discovering, on his arrival in Titusville, that he is in the very midst of the oil region ; the bar rooms of the hotels are crowded with mud spattered, travel stained men, who talk oil ; there is a rattle of glasses among rough clad, keen eyed men who have struck oil ; there is a Babel of tongues amid the cloud of tobacco smoke from all sorts of men ; and the words one can distinguish are "forty barrels a day—flowing well—seven hundred feet—sixty thousand dollars—oil—our company—bought at a bargain —sold—oil—got a new engine—run dry—new derricks—sold out and made—bust up and left—thousand dollars a day—oil—dollars—refine—down the creek—oil — another vein — oil — between you and I—oil—do not say a word—drill—d——d swindle— worth a million—oil—big thing "—and so on, till one's head almost swims with the greasy topic.

A barrel factory in this place turns out four hundred barrels per day, at three dollars and twenty-five cents per barrel. The Oil Creek Railroad is finished to this place from Corry, twenty-eight miles, and to the Shaffer farm, about seven miles down the creek, on its way to Oil City. The report that no railroad would be constructed below Titusville, through the heart of the oil district, on account of the danger of the gases taking fire from the sparks of locomotives, is believed to be a story set on foot by teamsters, who realize very largely by carrying oil from the wells to the railroad depots. The Oil Creek Railroad will be continued as fast as men can be obtained to build it.

This borough, from an humble country village only a few years ago, numbers now a population of some six thousand. New and handsome brick edifices and private dwellings are going up on every side, and indicate a determination on the part of the citizens to make it a substantial and permanent place of business. There are two banks here (the Petroleum and one National), and room for three or four more. The place contains thirteen hotels, and there is a fine opening for a first class house. It has a large hall, called "Crittenden," which will hold seven or eight hundred. It has one weekly newspaper. Among the residents are a number of New Bedford gentlemen—among them the brother and a son of the late mayor of that city, the Hon. Isaac C. Taber.

OIL CITY.

Oil City is about seven miles from Franklin, on the Alleghany river, at the mouth of Oil creek, which rises in the north-western part of Pennsylvania. For about a mile above Oil City, on the right hand side of the stream, the bank rises in an abrupt bluff, at the foot of which a very substantial road has been constructed on a stone foundation, over which teams are constantly passing, conveying oil to the river. The city itself is built at the base of a mountain, on the flats that run along the base of the high bluffs, and between them and the creek and river. It has but one street, and the grading of it has just commenced, and all the rocks, boards, boxes and rubbish generally are thrown into the middle of it.

The buildings on one side of the street all rest upon stilts or spiles, and occasionally one caves in, as the post office did the other evening. On the other side a man begins to build with a depth of first floor of twelve feet, the next twenty, the next thirty, according to the " perpendicularity " of the mountain. The population are all busy, like sensible people, attending to their own business and making money—but they go to church and close their grogshops on Sunday. The town is all wealth and mud—the creek all scows and scowling boatmen. When first built it was supposed, that the oil would not last, consequently, every person built his house in the cheapest possible manner, so that when the oil did give out he could leave, and

not be at much loss. The city has a temporary look
Directly acrosss the river and over the creek, on Cot-
tage Hill, a few fine cottages are going up, and they
present a pleasant appearance in contrast with the
wretched aspect of the " city " proper. Restaurants
and saloons, *par excellence*, are greatly resorted to, and
with the exception of the billiard rooms, there is no
approach to a decent room, in regard to size, in town.
The billiard room is quite a large and well-kept room,
having four tables up, and room for three more. Oil
City is the New York of the oil region. The whole
region known as " up the creek," is tributary to the
place. Steamboats from Pittsburg make regular
trips to Oil City—when the water will allow of it.
The many thousand wants of the community " up the
creek " are supplied from this point ; all the oil from
Oil Creek comes here to be overhauled, and is re-
shipped from here to the East ; most of the companies
having their offices here. Rooms are very scarce—in
fact there is a fortune for any one who will build a
large house here as a lodging-house.

The amount of business done by the banks is really
immense. Any day you can see men depositing from
ten to fifty and one hundred thousand dollars—in fact
ten thousand is a small deposit. The First National
Bank has five or six clerks, and does more business
than any two banks in Brooklyn. You meet here men
from every quarter of the United States. The oil
fever has spread over the land, and men flock to the
Mecca of their hopes in incredible numbers, and pay
fabulous prices for wells and territory, immense fortunes being realized in a few weeks. The price of
labor is very high, and laborers are very scarce. Adventurers, capitalists, broken down merchants, and
laborers from every direction, are attracted here, and
yet there is room for more. Oil-wells have become
greater fortunes than gold mines, and consequently

attracted the shrewdest, most plucky, and most venturesome business talent of the country.

CORRY CITY.

Two or three years ago, this place, which has now a population of about four thousand, was not in existence; and even at this day the city bears the signs of its reçent birth, the people not having taken time to remove from the front part of the place the stumps which they are obliged to dig out in order to clear a space for building. Three railroads centre here—the Philadelphia and Erie, Atlantic and Great Western, and the Oil Creek road. So fast have the people been to accumulate wealth, that as yet no churches have been completed. But there is a Catholic church, also a Baptist and Methodist under way. Eligible building lots command $300 to $800. Five years ago the whole site of the town might have been bought for the lower sum or less.

Mr. Bennett, the burgess, came here three years ago, and paid $2 to $2 50 per acre for land that now commands $700 to $800 per acre. Samuel Downer, of Boston, owns the extensive oil factory located in Corry, and rents it to the company that now carries it on. It is valued at $500,000. The works cost $175,000. They employ 175 men, and pay $1 75 to $3 00 per day. Have refined 100 barrels per day for the last month, consuming 240 barrels crude. The products of distilling are :—1, still gas ; 2, gasoline or naphtha ; 3, water separated ; 4, burning oil ; 5 lubricating oil, by chilling or pressing with ice, similar to the process in making linseed oil. Fifteen tons of ice are daily consumed in this process. The product of the oil region, from data obtained at this refinery has been about 5,000 barrels per day for the past year.

There are not many residences remarkable for architectural beauty, it is true, and much of the city appears to be somewhat of the mushroom order, but improvements are being made rapidly. The refinery is the largest brick building and the feature of the place. It is built and furnished in the most thorough manner, with all the modern appliances and improved machinery, cooper's shop, huge tanks for holding the oil, &c. One of the great oil tanks holds six thousand barrels of oil, and two others a thousand each. The product of the region, from data obtained at this refinery, has been about 5,000 barrels per day for the past year.

The Atlantic and Great Western, Philadelphia and Erie and Oil creek railroads, centre here, and twenty or thirty trains a day pass and leave, making it quite a stirring place, especially in the vicinity of the railroad station, where the vast train of oil barrels, the quantity of machinery, boilers, tubing, and pumping engines waiting transportation, and the rush, hurry and bustle, of an eager crowd, tells that we have reached the outer edge of the great whirlpool of treasure-seekers, mammon-followers and oil-borers. Like all new Yankee towns, Corry City has its newspaper, the Corry City *News*.

One of the handsomest wooden buildings is a large billiard hall, and the hotel is quite a tolerable one, and really as respects its table, is much better than some of the fashionable houses of the Western States, is called "The Boston House," its owner, Samuel Downer, Esq., of Boston, evidently believing that there is something in a good name.

Building lots in Corry command from three to twelve hundred dollars.

ROUSEVILLE.

A very large amount of businness is now transacted in Rouseville in the transfer of leases, buying and selling lands, &c. The country around abounds in rich oil territory—the Hammond well, on the Steele farm, being the most productive—a single well in the vicinity flowing two hundred and fifty barrels per day. The Trundy wells are also successfully worked.

The next farm of importance is the Rynd farm, lying between the Steele farm and Blood farm. It comprises three hundred and eighty acres of oil territory, on Oil Creek, at the mouth of Cherry Tree run, extending from the creek to Cherry run. Great activity is shown in the operations of this farm, there being some twenty-five paying wells working producing about one hundred barrels per day, and thirty in the process of boring. It is owned by the Rynd Farm Oil Company and judiciously superintended by Colonel Hoffman Atkinson, formerly Adjutant on General Smith's staff, Army of the Cumberland.

About a mile and a half above Rouseville is situated a little Island called Blood Island, on which several wells have been sunk, and which promises to be very productive territory. Below this point, the land seems better adapted to farming purposes than farther up the stream, sloping gradually to the water's edge, instead of rising in steep abrupt bluffs.

ALLEGHANY RIVER OIL COUNTRY, ABOVE OIL CREEK.

There are good wells on the Alleghany river, above Oil creek. The Wheeler well, and other wells all

3*

around to Walnut Bend, produce from ten to thirty barrels per day. Wells are being sunk on all the small runs—viz : Lamb's, Carey's, &c.,—up to their heads, embracing the flats between Oil creek and Walnut Bend.

Between Horse creek and Panther run, on the other side of the river, ten or twelve new wells are going down. One struck near Horse creek promises very good.

Coming up to Walnut Bend proper, the Continental Company, Philadelphia, have several good producing wells, and good property on the river. The Bruner Company, Philadelphia, have a good region, striking gas, which is a pretty sure indication of the presence of oil. The Star Company have one or two wells.

Pit Hole creek empties into the Alleghany. There are good wells about its mouth, and fair prospects up to Plummer road. All the territory above has been taken by companies, whose names are not well known. The creek runs up to Neilsburg, seventeen miles.

From the mouth of Pit Hole creek up to a place called President—where there is a good hotel, new, clean and plastered—there are wells going down on Harper's Farm. Hemlock creek empties into the river at President. Then comes the two Tionestas, Lower and Upper. Oil is produced at the mouth of each, but none at any distance from their mouth.

Next comes Hickory Creek, and thence we go to "Tideout," Warren county, where we find a sect called the "Economists," obtaining oil from shallow wells. Their settlement is on the river, and numbers about five hundred souls. They send their oil out by way of Irvin, on the Erie and Philadelphia road.

This is the head of the oil region on the Alleghany. Next, going down below the mouth of Oil creek, looking after the branches, is the Big Sandy, the borders of which are all bought or leased for oil territory for a

long distance from its mouth. Scrub creek is next in
order ; but no oil of much account has thus far been
found on either stream. Opposite from these streams
is the East Sandy . but no oil has been discovered
there.

Two miles above Franklin Two Mile run empties
into the Alleghany, and about two miles above its
mouth a small amount of oil of very good quality has
been discovered. All the territory has been leased,
and is considered valuable, as it runs parallel with
Oil creek after it gets up a short distance.

ALLEGHANY RIVER OIL DISTRICT BELOW FRANKLIN.

Messrs. Dale & Morrow struck a vein a short time ago
on the Cochran Farm, two miles below Franklin, which
yielded 240 barrels the first forty-eight hours pumping. It is considered good for 100 barrels a day.
John Lee, of this place, has also obtained a flowing
well on the Martin Farm, just above the Hoover, and
nearly opposite the Cochran Farm, which flows over
50 barrels daily. Besides this, a well has been struck
on the island opposite the Hoover wells, which promises to be a first class well. It has not been tubed
yet.

In addition to this, Mrs. Hubbs, who owns a lease on
the Smith Farm, four miles below Franklin, now the
property of the *Excelsior* Oil Company, and who had
been pumping but four barrels a day at the depth of
408 feet, sunk his well to the depth of 424 feet and
struck a vein which produces 40 barrels in twenty-
four hours.

Also, we note the striking of two good wells on
French Creek, one and a half miles from this place.
All the above strikes have been made within a week,
and there are good prospects of a number more in a

short time. A well has been opened on the Brown
Farm, on Sugar Creek, near its junction with French
Creek, which is yielding twelve barrels per day.
Several others are going down in that vicinity with
very flattering prospects. The consequence is, that
territory in that vicinity is going up in price. There
is one fact which oil operators appear to overlook, and
that is that there has been many successes in proportion to the number of wells bored, on French Creek
and its tributaries, as on Oil Creek and its branches.

The well recently opened on the Hoover Island a
few miles below Franklin, is now flowing about 100
barrels a day and is increasing.

Oil territory on the river below this place and on
French Creek, is being taken up fast at high figures
and in most cases will prove valuable investments to
the purchaser.

We are informed that Messrs. Morrow & Dale, have
sold one half the working interest in their well recently opened on the Cochran Farm, for $50,000.

The first well one comes to, on leaving Franklin, is
one belonging to the " River Oil Company," which is
now 750 feet deep, and still boring. A good show of
oil was obtained at 700 feet. Just opposite, on the
other side of the river, is the " Faulkner well," pumping three or four barrels of oil per day, 360 feet
deep. Next are the Keystone wells. One is 460
feet deep, and pumping some four barrels per day.
Next is Cochran's and Williams' well, 430 feet, boring
with an excellent show of oil. Just across the river
are two wells with good shows, styled the " Spragle
wells." Just below, on the same side of the river, is
the " Lee well," depth 489 feet, and flowing sixty barrels per day, as near as it is possible to estimate.
This company have three other wells. The next is
the " Dale and Morrow well," above mentioned, situated on the Cochran farm, about 440 feet deep, and

pumping about forty barrels per day. Another well belonging to the same company is producing six to eight barrels per day, about 450 feet deep. The two wells were bored with one engine.

Next is the " Shippen well," 460 feet deep, with good show. The next well is on Plummer and Hoover's Island, the first island below Franklin. It is 427 feet deep, and flowing. When pumping, it will produce about fifty barrels per day. The next works belong to the " Pennsylvania Coal Oil Company," who have four wells in various stages of progress. On the opposite side of the river is the " Reinhard wells," producing twenty barrels per day. Just adjoining is the " Hoover and Marshal" well, a good pumping well, and next to this are the works of the " Roberts' Oil and Mining Company." They have one well producing eight barrels per day, and four wells nearly completed.

Next is the property of the " Greenville Oil Company," which have one well producing some oil ; also the wells of General J. K. Moorehead, one of which is producing. Next is the " Cranberry well," being now 546 feet deep. On the adjoining lands of the " Alleghany River Coal Oil Company," is the well of the " Smoky City Oil Company," which has been producing for three years, pumping night and day, with no evidence of exhaustion. The " Hope Company " have a well on this land, but they have made arrangements to bore for the third sand rock, to test the theory, that if the third sandstone is reached, which is to be found at about a depth of a thousand to twelve hundred feet, immense flowing wells will be the result. The " Alleghany River Coal Oil Company " have two wells going down ; one of them is named the " Kendrick," and the other the " Weaver."

FROM PETROLEUM CENTRE TO SHAFFER.

The town of Petroleum Centre is situated on the west side of Oil Creek, and is so called from being about an equal distance from Oil City and Titusville. It is a collection of about thirty dwellings, a fine hotel, a dozen shops and stores, a telegraph office, post-office, &c. It has, moreover, the advantage of a bridge over Oil Creek, of which there are but three between Oil City and Titusville. The property adjacent on both sides of the Creek is known as the George W. McClintock farm, and adjoins the Egbert farm on the south, the McRea farm lying to the east of both on the hill. On the north is Bennehoof Run and the Boyd farm, and immediately adjacent to the town, what is known as Wild Cat Run, a circular ravine of about three-fourths of a mile in length, running around a "hog-back" of some fifty feet elevation. The farm contains 255 acres, and is owned by the Central Petroleum Oil Company of New York. There has been no considerable effort towards the development of this property until the past year, and there are now eighteen wells producing, yielding together about 550 barrels per day. Of these four are flowing and the others pumping ; ten are operated upon leases which pay half the product to the land interest, and eight are worked by the company. There were but thirteen leases on this property at the time of its purchase in March last, and the company have determined to execute no others, but rather to develop it themselves, and for this purpose are now preparing to sink a number of additional wells. The first well put down on this farm, commenced flowing in August, 1861, at a depth of 472 feet, yielding about 500 barrels per day, but ceased in December, after flowing about 50,000 barrels. Another well opened in November, 1861, at a depth of 440 feet, flowed about 100 barrels per day for a time, but then ceased. In January, 1862, it was deepened fifty feet, and afterwards flowed about forty barrels per day. It is now pumping ten barrels. The wells since put down average about 500 feet. The Coldwater wells (two) are yielding about forty barrels Wild Cat Run is owned chiefly by the Petroleum Centre Company of Philadelphia. Quite a number of wells have been put down in the Run, but only four or five are producing oil, their yield being from ten to thirty barrels per day. The Sherman Oil Company of Philadelphia own half the working interest in the Gillespie tract on this Run, on which is one well producing and another in process of boring. The same company also have the land interest in the Swamp Angel lease on the McClintock farm, now yielding fifty barrels per day. On the elevated ground which forms the McRea farm, several wells are in progress, the belief being that here as at various other points on the Creek, the oil veins pass over the hill eastwardly, whatever may be the occasion of the unproductiveness of like situations on the western side of the stream. In Bennehoof Run several wells are in progress, and three or four old wells, formerly yielding from ten to twelve barrels per day, are now idle. The Pennybaker well, in which the Mingo Oil Company of Philadelphia have a full fourth interest, is now yielding fifty barrels per day. Near the mouth of the Run a new well has just been put down to the depth of 600 feet and is now being tubed. The Boyd farm adjacent embraces about seventy-five acres, and is now owned chiefly by Messrs. Wood & Wright of New York, who are making active efforts to develop it. Fifteen wells were put down on the property in 1861 but of these only one was put in operation, yielding from ten to twenty barrels per day. Three wells put down during the summer are now producing, yielding together about thirty-five barrels per day, and four or five additional wells are now in progress. The Patterson well, deepened during the summer to 500 feet is yielding about ten barrels. The Brownsville Company's wells, and also the old Snyder well are idle.

The McIlhenny or Funk farm, embraces about 180 acres, and is owned one-third by the McIlhenny Oil Company, one-third by the Dalzell Petroleum Company, and one-third by other parties, each of the three interests being entitled to a sixth of all the oil produced. It is a point of considerable interest, a small town, called Funkville, having sprung up on the upper section of the lower farm, while the flat below is almost literally covered by derricks, engine houses and tanks. This property, then owned by Captain Funk, was largely productive in 1861 and 1862, several of the first wells opened on it flowing from 200 to 2500 barrels per day, at a depth of from 450 to 475 feet, which is about the average of the wells since put down. The Crocker well, which was the first one put down, commenced flowing in January, 1861, and flowed 1000 barrels per day for some time, but afterwards rapidly fell off, and in 1862 ceased altogether. In September, 1861, the Empire well commenced flowing about 2500 barrels, and yielded 2000 barrels per day for most of the winter, falling off during the summer to about 300 barrels, and ceasing in April, 1863, after flowing 80,000 barrels. The Buckeye well commenced in September, 1861, flowing 800 barrels per day, fell off to 200 in 1862, and ceased after flowing 43,000 bbls. The Fertig and Funk wells also opened in 1861, flowed 200 bbls. each daily, and ceased in 1862, after yielding about 40,000 bbls. each. The Burtis, Aiken, Davis and other wells were at the same time flowing from 15 to 30 barrels per day. All these wells have since been idle, with the exception of the Empire, into which a patent air condenser was introduced in July last, and the well has since produced from thirty to sixty barrels per day, but is now again idle in consequence of some injury to the machinery. A like apparatus has been put into the Empire well, No 2, formerly yielding 150 barrels per day, but for some months idle, and it is now yielding twenty barrels per day. A third has been introduced into the Buckeye well, owned by the Briggs Oil Company, but as yet has had no beneficial result. The present product of the farm is about 320 barrels per day, being less than at any time during the last year. The Olmstead well, owned by the Olmstead Oil Company of Philadelphia, and formerly yielding 300 barrels per day, has been yielding nothing for the last two months. The company have, however, an acre of land on which they are about putting down other wells. The Hibbard wells, Nos. 1 and 2, owned by the Hibbard Oil Company of Philadelphia, are also idle, one of the wells having exploded its gas tank about the first of December last, and not yet resumed work. The Forest City, yielding until recently forty-five barrels per day, is now doing nothing ; the Dinsmore, No. 6, is yielding 100 barrels ; the Hatch well, No. 3, fifty barrels ; the Fertig, ten ; the Harding, Mount Vernon, Long, Dean, Wilson and other wells, are yielding from ten to twenty barrels. There are about fifty abandoned wells on this property, including the lower and upper farms, and a dozen or more new wells are in progress.

The Foster farm adjoins what is called the upper McIlhenny, lying to the west of the creek, and here is situated the well known Sherman well, put in operation in March, 1862, and for some months one of the largest flowing wells on the creek. It commenced with a flow of about 2000 barrels per day, but after a few months ran down to 600 barrels, and ceased flowing in February, 1864, since which time is has been pumped, yielding during the summer about 40 barrels per day, but stopping altogether early in October. An air condenser has now been put in the well, and it is again yielding from ten to fifteen barrels per day. The Foster, Dale, Gordon, and other wells, formerly yielding from 50 to 100 barrels per day, are now idle. Immediately above the Sherman well lies the property of the Irwin Oil Company of Philadelphia, on which is the Irwin well, formerly yielding 120 barrels per day, but which gave out during the last summer. The Irwin Company have from twenty-five to thirty acres of land, with a front upon the creek of about 500 feet, and running back diagonally over the hill to a small ravine beyond, in which they have a well in progress. In the same locality is the Pioneer Refinery, and still other wells have been projected by different parties on the ravine. The company have a new well near the creek front that has yielded some oil, and an old well a few rods back, near the Irwin, has been deepened to 500 feet, and is now about being put in operation, new engines having just been procured both for this and the new well. Near the Irwin Company's property is the Porter well, which commenced flowing in May last, yielding during the summer from 70 to 150 barrels per day, and now pumping about 20 barrels. A number of other wells on this farm are pumping from five to fifteen barrels per day, and several new wells are in progress.

The Farel farm, on which is situated the Noble well, is immediately opposite on the east side of the creek, and is mostly bluff. There are here eight or ten wells, but only the Noble and the Milligan are producing, the former flowing about 200 barrels per day, and the latter 30 barrels. Three new wells put down during the summer have as yet produced but little oil. The Noble well commenced in May, 1863, and for the first three months flowed about 1700 barrels per day. During last summer it was yielding about 600 barrels per day, but has latterly ran down to 100 a 200 barrels, the oil also being somewhat rily. The well is 470 feet deep, and has yielded up to the present time about 410,000 barrels of oil, selling for over $2,500,-000. It is owned, together with some fifteen acres of land upon the bluff, by various companies and parties. The Noble and Delameter Petroleum Company own $\frac{49}{100}$ of the whole ; the Noble Well Oil Company one-sixteenth, and the balance of the interests are held in various proportions from one twenty-fourth to one ninety-second, by the Farel Oil Company, the Union Petroleum Company, Wood & Wright, V. M. Thompson, John Farel, Lamb, Churchill & Emmon, Jefferson, M. Wilcox, J. W. Hammond, J. H. Elmore, S. F Dewey, Everett & Russell, Charles Delameter & Co., Noble & Hale, and the Success Oil Company. The Caldwell well, on the bluff below the Noble, which with one acre of land was purchased by the Noble Company for $143,000, has not since been operated, its purchase having been a measure of protection merely. A new well recently put down by the company in Bull Run, a short distance above the Noble well, is believed to have interfered with it, and has not for that cause been worked.

There is no considerable production of oil at present above the Foster farm. Upon the several flats or farms are numerous abandoned wells, and at various points quite a number of new wells are in progress. On the Gregg farm, immediately above the Foster, the Prentiss well, which has been in operation four or five months, is yielding from 10 to 15 barrels daily. This well was sunk over a year ago, and for a time yielded about 30 barrels. The Wilmington well formerly yielding 50 barrels, is again about starting The Sloan well (new) has been sunk to the depth of five hundred and sixty feet, with no indications of oil. The White well is yielding about thirty barrels. The first well sunk on this part of the creek, between the

Buchanan farm and the old Drake well, a mile and a half below Titusville, was the old McCoy well on the upper point of the Gregg farm. It was sunk in 1859, and at a depth of one hundred and eighty-three feet yielded from ten to twenty-five barrels of oil per day from March until July, when it was burned out. It was afterwards run down to five hundred feet, but has yielded nothing at that depth. Immediately above this is a well, sunk last summer, that at six hundred feet has afforded no indication of the presence of oil. On Sanny farm above, and to the east of the creek, are a number of abandoned wells, some of which also were put down to the depth of six hundred feet without finding oil, or reaching the third sand-rock. Two of these are now being deepened.

The Shaffer farm, ten miles from Oil City, and seven from Titusville, is the present terminus of the Oil Creek Railroad, and within the past six months has become a place of much importance. The farm lies on the west bank of the creek, opposite a sharp bluff, the flat being about 200 yards in width, and from the rear of which the hill rises gently to the westward. There had been a number of wells put down on this property three years ago, some being 450 and others 500 to 550 feet in depth, but none ever producing much oil. On the upper point of the flat are two wells owned by Samuel Downer, one of which, the "Rangoon," was opened in 1863, and is now pumping 15 barrels; the Railroad well, adjacent, commenced in April, 1864, yielding 45 barrels per day, but for the past two months has not been in operation. On the east side of the creek, Messrs. Brewer & Watson have a well yielding 30 barrels per day. The same parties have other wells adjacent, one of which, in 1863, yielded upwards of 1000 barrels per day for a considerable time, and another about 100 barrels per day, but none of these are now in operation. There are on this property several leases, and quite a number of wells are projected, particularly on the lower section of the flat, but none are yet producing. Messrs. R. Wildy and A. C. Roberts have a well now being put in operation, that at 510 feet depth, affords the most encouraging indications. The great feature of the place, however, is the railroad depot, and the very thrifty town which has grown up within the past six months. Previous to August, 1864, the shipping point on the Oil Creek Railroad was at the Miller farm, a mile above. In July the track was completed to Shaffer, the road running along the edge of the hill, at the back of the flat, where commodious landings, depot buildings, turnouts, &c., have been constructed. The depot buildings occupy the western side of the main track, the landings in front forming an elongated semicircle of perhaps 1000 feet in length, around which there is a double track, lined on the creek side by a number of spacious wharehouses.

The shipments of oil from this point have been large, averaging since August last, about 2000 barrels per day, and the business of the road other than this, has been immense, every train bringing down vast quantities of machinery, iron, tools, pipes, tubing, lumber, barrels, coal, &c., &c., so that the landings, capacious as they are, are kept constantly crowded. It is no unusual occurrence to see thirty or forty engines on those landings at a time, and dozens of car loads of tubing and other machinery awaiting transportation, down the Creek, while at Titusville and Corry the road is literally blocked with freight. In the meantime the town of Shaffer has grown rapidly. In July last, when the railroad buildings were commenced, the place was marked by a single frame dwelling, and a small restaurant on the edge of the Creek near the Railroad and Rangoon wells. Now it can boast four excellent hotels, a dozen or more stores, two extensive livery stables, a post-office, some fifty dwellings, and an excellent school. The most commodious hotel in the place is the Person's House, capable of accommodating one to two hundred guests. The others are the Shaffer House, Shaw's Hotel, and the Jamestown House ; and the tide of travel is such that all are constantly crowded. The buildings are all frame, and mostly of inexpensive construction. The cost of living is high, taking into consideration the fare provided ; the charges at the hotels being from $2 50 to $3 per day. Everything in the way of marketing has to be brought from Titusville, Meadville, and Salamanca. The livery stables are doing a large and flourishing business, each keeping from fifty to one hundred horses, for which there is a constant demand at from four to five dollars per day. The population of Shaffer is about 400, exclusive of strangers and the numerous persons employed about the depot in various capacities.

The greater portion of the oil produced on the Creek, as far down as the Tarr farm, usually comes to Shaffer for shipment, the charges for hauling northward to Shaffer, or southward to Oil City, being generally about $1 50 per barrel, but varying much according to the condition of the roads. A two-horse team will ordinarily haul six barrels, returning with a load of empty barrels, which are carried at a charge of twenty cents each. When, however, the Creek is navigable for boats, and on the occasion of pond-freshets, the great bulk of the product, even as far up as the Foster and Farrel farms, is sent down the Creek for shipment. The boats are usually towed up the Creek, the charges for towing empty boats of the capacity of 150 or 200 barrels ranging from $20 to $40, according to distance. When towing is practicable regular communication between Oil City and Shaffer is kept up by this means. Hundreds of travellers visiting Oil City take the railroad to Shaffer, and there embark in open boats, often without seats of any kind, for Oil City, or any intermediate points they may desire. The fare for such conveyance is modest, being only about fifty cents per mile, say $4 to Oil City. Much freight is also taken down the Creek by this means, when it is impossible to convey it by wagons in consequence of the condition of the roads. Such is the difficulty, at times, of moving heavy freight, that not unfrequently fifty and one hundred dollars have been paid for transporting an engine a single half mile.

A large business in coal is growing up at Shaffer. Most of the coal used on the creek has heretofore been wagoned from the Cranberry coal banks, some seven miles southeast of Oil City. In the summer coal is supplied at the wells at from 25 to 40 cents per bushel; but such was the condition of the roads during the months of November and December last, that prices advanced to $1 30 and $1 50 per bushel; or $30 to $35 per ton, and operations at numerous wells were suspended for the want of coal. Coal of a much better quality is now found on the line of the Philadelphia and Erie Railroad, in the vicinity of Ridgway, and is being shipped to Shaffer, and thence on boats down the creek, at prices which yield a large profit. At some of the wells the custom is to burn the gas from the well for the purpose of driving the engine. A pipe is put down through the seed bag by the side of the tubing, leading from thence into a hogshead half filled with water, where it undergoes purification, and is then conducted by another pipe to the furnace. It is now believed, however, that such use of the gas causes more injury to the well by diminishing its yield, than is saved in the cost of fuel, besides at times being attended with disastrous results. Any material diminution in the supply of gas, even though but momentary, leads the flame directly into the tank, resulting in immediate explosion, and sometimes heavy loss from fire. It is only a few weeks since the Hibbard wells, and a considerable stock of oil, were imminently endangered by an occurrence of this character.

The Oil Creek Railroad, which is now under the joint control of the Philadelphia and Erie and the New York and Erie Railroad Companies, is being rapidly extended down the creek, the grading for the road bed, which follows the hill to the west of the creek, being nearly completed as far down as the McIlhenny farm. The completion of this road will add very greatly to the facilities of transportation, as well of oil as of machinery and fuel, and will open a new era on the creek. The present means of transport are of the most primitive character, and are not only costly but attended with many difficulties.

In the fall and winter of 1863, after the Oil Creek railroad had been opened to the Miller farm, a number of gentlemen formed themselves into a company, for the purpose of pumping oil the whole distance of the creek, by means of iron pipes. They contracted for about ten miles of iron pipes, of about four inches diameter, most of which was delivered on the creek, and about three miles put down, extending from the Miller farm to a point on the Foster farm, opposite the Noble well; but the enterprise could not be made to work successfully, and was abandoned, all the parties concerned in it losing largely.

The pumping was done by a stationary engine, the oil being supplied from a tank capable of holding five hundred barrels; but in consequence of leakages in the pipes, there was a large deficit in the deliveries at the upper end of the works, the loss in many cases, amounting to ten per cent., being very nearly, if not quite, the cost of hauling. Dealers were unwilling to

lose, and the company unwilling to make up this lose
amounting to double their charges, and in consequence
the whole project went by the board, The pipes still
line the creek at various points, bearing silent testimony
to the disappointed hopes of the parties engaged in
the enterprise.

OIL CREEK, ABOVE ROUSEVILLE.

The connection between different wells, entering
the same oil veins, which tends so largely to the inse
curity of leased property as well as small land interests
in the Oil Creek Valley, has nowhere been more stri-
kingly illustrated than on the farm formerly owned by
the Widow McClintock, and by her bequeathed to Mr.
John Steele. It is immediately above and on the opposite side of the creek from Rouseville, and contains
about two hundred acres, of which nearly one half is
bottom land, with a frontage on the creek of about half
a mile. There are upon the farm upwards of twenty
producing wells, yielding about two hundred and fifty
barrels of oil per day, besides some forty abandoned
wells, and ten or twelve new wells in process of boring.
Some of the most productive wells on this property
have been located upon the bank of the creek where
the Van Slyke well, put down in 1861 to the depth
of 640 feet, flowed from 1000 to 1200 barrels per
day ; the Chirty 30 barrels, and the Eastman, Hayes
& Merrick and other wells, from 20 to 50 barrels per
day. Latterly, however, these wells, located in almost a direct line, have produced no considerable
amount of oil. The Van Slyke, which had produced
an aggregate of 40,000 barrels, was in May last yielding about 50 barrels per day as a pumping well ; the
Lloyd 20 barrels, and the Christy six or eight barrels,
the others being idle. About the middle of May the

4

" Hammond" well, which had been put down by certain New York parties to a depth of 600 feet, directly between the Christy and Van Slyke, Lloyd and Christy produced but very little oil, the daily yield ranging from three to ten barrels. The Hammond continued to flow until about the 10th of June, when together with nearly a dozen other wells upon the flat, it was flooded by the drawing of the tubing of the Excelsior well on the John McClintock property, about 500 yards distant on the east side of the creek, and has since produced very little oil, except for a few weeks during the summer, when the Excelsior having been retubed, the Hammond yielded for a time, as a pumping well, about 150 barrels per day. Early in June, and but a few days before the Hammond was first flooded, the land interest in the well, including less than one-third of an acre, was purchased by the parties owning the working interest for $200,000.

When these wells were first flooded, it was supposed to be the result of sinking a new well, a short distance below, on the flat, by Messrs. Vandegrift and Titus ; but careful observation soon demonstrated the connection existing between all these wells, and that the Excelsior well was the source of the supply of the water which all were pumping. It appeared, moreover, that the Excelsior had ceased yielding oil within a few hours after the Hammond commenced flowing, and further, that the pumping of the wells vpon the flat had daily a perceptible influence upon the water in the Excelsior. After much difficulty, the owners of the two wells came to an understanding during the summer, by which the Excelsior was retubed, and the water being thus shut off, the Hammond produced for a short time 150 barrels per day, bvt the other wells adjacent have never been restored. Subsequently the tubing was again withdrawn from the Excelsior, and the Hammond has since pumped nothing but water. In numerous other cases the connection between wells is proven with equal clearness, but the facts connected with the Hammond well raise another question of no little interest. The well was put down between two others already nearly exhausted. Subsequent facts prove that it entered the same veins as these wells, and being no deeper than these, its large production, while the others were yielding little, can only be accounted for by the supposition that it penetrated a vertical vein below the range of the horizontal veins supplying the adjacent wells. Such being the case, it would be liable to be flooded with the others, by the introduction of water into the horizontal veins, which being again shut off, as was shown by the retubing of the Excelsior, the well would resume its flow of oil wholly independent of the wells adjacent. In any case, the question is one of much interest, and it is not doubted by the owners of this well that it will yet become a valuable property, whenever they shall be able to bring their difficulties with the owners of the Excelsior to a proper adjustment. There are two new wells upon the Steele property, producing about 40 barrels each ; the two Painter wells are yielding together about 50 barrels ; the McClintock 10 barrels ; the Morrison & Bell 20 barrels ; and several others from 5 to 15 barrels.

PETROLEUM IN OTHER PLACES.

The Morgan Herald reports that a well, offering a good yield of oil, has been obtained by boring two hundred feet, within two miles of McConnellsville, on the Kirkpatrick farm.

New Strikes on Sandy Creek.—We learn, says the Oil City *Monitor*, that the Keystone Oil Company, Capt. Pasely, Super intendent, struck a splendid well on Wednesday of this week The well is not yet tubed, but it is variously estimated at from forty to sixty barrels. This well is about sixty rods below the Slate Furnace. The well at the mouth of the creek, struck about ten days ago, continues its yield as heretofore. This, we think, demonstrates the position long since maintained, that East Sandy Creek would, when sufficiently developed, prove to be first class oil territory. Months before the great oil excitement broke out, we greased our boots with the oil from the natural oil springs at Hall's Run. For years we have been in the habit of visiting occasionally this spot with curiosity seekers to see the oil spout up through the water from the crevices in the rock. As we have before stated, this spring, which yields its unfailing supply of oil, proves not only the prevalence of oil in large quantities, but also the open structure of the rock, a condition quite essential to permanent wells. We expect soon to hear of more than one Cherry Run and Oil Creek in our county, and East Sandy and its tributaries bid fair to come in for a large share of consideration.

The Ritchie Well.—The following item from the Oil City *Monitor* will be read with interest by the stockholders in the Ritchie Oil Company, recently organized. Our cotemporary says: "The Williams well, on the Clapp farm, is pumping 150 barrels. It was sold to Claney, Tyler & Co., and since put into a stock company, known as the Ritchie Oil Company, formed on a basis of 160,000 shares at $1, and on this capital the proceeds of the working interest of the well being from seventy to seventy-five barrels of oil per day, and selling as it now is, at from $11 to $11 50 per barrel at the well, it will pay from ten to fourteen per cent per month. There was an offer made recently for the Ritchie interest of this well at $200,000, or test the

well and pay $3,500 per barrel. We consider it the best founda-
tion for a company we have heard of for some time."

Coal Oil in Jasper county, Ind.—There is some excitement in
Jasper county on the coal oil question. Petroleum has been
discovered in the ledge of rocks upon which the town of
Rensselaer is built, and the same characteristics pervade the
entire line of the Iroquois rapids. Two companies have been
formed, and headed by Mr. Brown, the State Geologist, and
backed up by parties who have experience in the oil business of
Pennsylvania. The indications of oil are undoubted and parties
interested are sanguine of the best results. Land in Jasper
which has hitherto been rather a drug in the market is looking
up, and prices have been greatly stimulated.

A Valuable Well.—An oil well near Zanesville, Ohio, is now
yielding 160 barrels of oil a day and the oil sells at the well for
$24 a barrel.

Oil at Fishkill.—We have been shown two specimens of Fishkill coal oil, gathered from springs. It has the *scent* of the real
article, and perhaps shows where it may be found in large quantities. In the meantime the Fishkill Standard says a company
has been formed with a capital stock of $600,000, all of which
has been taken. The matter has been taken in hand by the
most prominent, wealthy and enterprising men of that vicinity,
and operations are to be immediately inaugurated for the get
ting of oil. It adds that the indications are favorable, and numbers of gentlemen, who were skeptical at first, have had all their
doubts removed by a visit to the " oil regions," and have taken
stock in the company just formed.

Oil in Colorado.—The Denver Daily News, of October 28
speaking of the oil wells of Canon city, says—

" Our people are, perhaps, not generally aware that Colorado can produce her own oil ; not only sufficient for home consumption, but a full supply for the neighboring territories. Good indications of coal or rock oil have been found near Canon city. These springs have been improved to some extent, but not sufficiently to yield any very considerable quantity of oil. The natural flow from the surface is about one barrel per day, and it is reasonable to suppose that the quantity can be indefinitely increased by judicious and extensive borings. We believe that in no other part of the United States are the surface indications as good as at the springs named. In the rich oil regions of Ohio and Pennsylvania we believe that surface indications are very rae. It is found by deep borings, and seldom, if ever, flows from natural springs. The fact that we have here a large oil spring would seem to indicate an unusual quantity of it beneath.

Petroleum in Stark county, Ohio.— The " oil fever " is raging in Stark county. The " Canton Oil Company," composed of speculators from Pittsburgh, has purchased six hundred acres of territory at Canton and on Bull Creek, and is sinking wells upon it. A rich flowing oil well is reported to have been struck a half a mile east of Waynesburgh, in the same county.

Oil in Tulpehocken, Ohio.—We learn that the people of Tulpehocken are excited at the report that oil had been discovered on the banks of Little Twin, in the vicinity of " big nose John Emrick's." It is said to be a very fine article of cod liver oil.

The Clearfield county [Ohio] Republican says : " Our people are taking the infection. Politics, the war, drafts, conscriptions, high taxes, and even our lumber business, all are in danger of being forgotten in the rage for petroleum," but so far nary " smell" of "ile" has been struck.

HE HERALD'S SPECIAL REPORT ON THE OIL REGION.

I have just returned to this pleasant town from a hasty expedition into the Oil Creek region, and will give you a few views in regard to the country, its population, &c.

THE EARLY HISTORY OF THE DISCOVERY

has not been fully understood, and a few remarks in addition to a former account may not be unacceptable. A Gazetteer of Pennsylvania, published as early as 1803, refers to the existence of oil in the western and northern parts of Pennsylvania by relating the fact that a noted doctor was in the habit of riding over the mountains on horseback, taking his saddle bags filled with bottles; then filling the bottles with the oil taken from the ground would return and sell it to the primitive settlements east of the Alleghanies, as a panacea for all the ills that afflict mankind. Although the existence of the oil was known more than a century ago, it was not until within the last twelve years that any effort was made to make it a staple product of the country, and send it to all parts of the world. Its development and its extensive use, not only in this country but in Europe, where, among other services, it is now used to light the streets of St. Petersburg, and in South America, where it furnishes light to illuminate the apartments of Brazilian princesses, and the more remote parts of the world, such as Asia and Africa, whither it is an article of ex

tensive export from this country, may be looked upon
with astonishment and incredulity by the casual ob-
server ; but those who carefully examine all the facts
connected with the subject can but wonder that it was
so long allowed to remain undeveloped as an article of
general use and commerce.

INCIDENTS OF ITS EARLY DISCOVERY.

In the spring of 1853 Mr. George H. Bissell, formerly of New Orleans, upon his return North from
that city, saw at the office of Professor D. Crosby, of
Dartmouth College, N. H., a small bottle of petroleum
given Dr. C. by his nephew, Dr. Brewer, of Titusville,
Pa.   Mr. Bissell became greatly interested in the product ; and about six months afterwards associated
himself with a Mr. Eveleth in developing the article.
These gentlemen proceeded to Titusville, and purchased, what was then considered the principal oil
lands of Pennsylvania, from Messrs. Brewer, Watson,
& Co., of Titusville, for $5,000.   Messrs. E. & B.
then organized a company, under the name of the
" Pennsylvania Rock Oil Company," and proceeded to
develop the lands by trenching them and raising the
surface oil and water into vats.   These primitive operations were conducted for about three years, only a
limited amount of oil, however, being raised, and that
used principally by liniment makers.   In the spring of
1855 Messrs. Eveleth & Bissell, at considerable expense, employed Professor Silliman, of Yale College,
to analyze the oil, and furnished him with all needful
apparatus for his experiments.   Professor S. was engaged about four months in his analysis, and in the fall
of 1855 Messrs. E. & B. published Professor Silliman's very full and elaborate report.   This report excited some attention in New Haven, and some gentlemen proposed to Messrs. E. & B. to re-organize in

New Haven. This was done, Professor Silliman being the first president of the company. The work of
raising the oil by trenching was continued until 1858,
when an arrangement was concluded by the company
with some of its members to bore an artesian well on
the land. A new company was then formed, called
the Seneca Oil Company, who obtained from the Pennsylvania Rock Oil Company a lease of the land for
forty-five years, on condition of giving the latter company twelve cents per gallon for all oil produced.
They employed a gentleman by the name of Drake—
then a conductor on the New York and New Haven
Railroad—to oversee operations on Oil creek, and
furnished Mr. Drake with the necessary capital. Mr.
D. commenced operations at once ; but was delayed by
many obstacles, until, finally, on the 28th of August,
1859, he " struck ile " at a depth of only sixty-nine
feet from the surface. This well produced about four
hundred gallons of oil per day, which, at that time,
was regarded as a " big thing." The excitement soon
spread, and from this simple commencement the
petroleum business—now the second article of export
from the United States, the production amounting since
January to sixty millions of gallons—was ushered into
existence.

THE QUALITY AND GRAVITY OF THE OIL FOUND ON OIL CREEK, ALLEGHANY RIVER, FRENCH CREEK, ETC.

It is interesting to those who are enlisted in the
business of developing this remarkable article to
know the gravity of the oil as found in different localities, as upon that depends its value for particular purposes. We will state, therefore, that the oil found on
Oil creek varies in gravity from forty to fifty degrees
Beaume. It contains a large proportion of benzine.
The oil of Alleghany river ranges from thirty-four to

thirty-nine degrees and contains very little naphtha. Much of this oil if mixed with lard and other substances, makes a very good lubricator for coarse machinery. It is, however, quite equal to the oil of Oil creek for illuminating purposes. The oil of French creek ranges from thirty to thirty-one and a half degrees, and is probably the finest lubricator known. It is used with great success on the most delicate machinery. This oil sells readily for about twenty dollars per barrel at the wells, not inclusive of package, and the price has remained the same for the last year and a half. Active operations are going on upon French creek, and the territory is being rapidly developed.

STEAM TO OIL CITY.

The roads from this point to Oil City being too bad for the carrying of freight, although the distance is but seven miles, a couple of snug little steamers, called the Petroleum No. 2 and Advance No. 2, have been placed on the river route, and are doing a lively business in the transporting of freight and passengers.

OIL CITY.

A subscription of fifty thousand to sixty thousand dollars has been started in Oil City for the purpose of building a new hotel there—a much needed improvement: and we are sure it will not be allowed to slumber by the oil nabobs who abound in that locality, and are celebrated for their public spirit.

OIL CITY AND ANECDOTES—ITS NOTABLES.

We come now to one of the most unique and interesting portions of the oil region. I find here a most extraordinary amount of mud everywhere; but every body is in good humor. The early history of Oil City may not be uninteresting. In 1860 a number of enterprising youths met at the Petroleum Hotel, then kept by one Colgin, and concluded to call the place Oil City. It had previously been known as Cornplanter village. The landlord of the Petroleum was an original, who had his own ideas about keeping a hotel. He used to say that "if the hotel was in good order he thought he could run her." Crapo, of New Bedford, was at this time at Tideout, from which he removed to Rouseville, and thence to Oil City, where he is now running the Crapo House. The Sheriff and other hotels have since been started. The opinion that where there is always so much mud there is no use keeping a nice hotel is fallacious. If the oil citizens would only "mend their ways" there would be no difficulty in keeping a tidy hotel, for there is money enough laying around loose there to sustain a first class house. Money is no object to these people, unless you come across a very close one. I occupied a double bed the other night in a room in which there were five or six others. About midnight one of the gentlemanly hotel assistants ushered in a stranger, with the remark that he knew us and desired to share part of our couch. We regarded the stranger for a moment, and hesitated to comply with the modest request. "Well, mister," said our friend, with a gentle wave of his body to and fro, "if you don't give me part of your bed, you needn't; but there isn't any other place for me to sleep but the sidewalk, and (hic) there isn't any sidewalk." We concluded to allow him a part of the bed; but he had scarcely touched the pillow when he commenced snoring as if he intended to start the roof from the rafters. We gently punched him in the ribs, and he uttered, in a beseeching tone of voice, "Mister, do you want a thousand

dollars?" We replied we had money enough—only
desired him to stop snoring. "Well, mister, if you
do want a thousand dollars you shall have it, by gorry."
It is thus with some of the natives of this curious
region. While some, we say, are avaricious and are
continually grasping for more, others again are liberal
to the last degree, and no worthy object of charity is
allowed to pass unbefriended by them. Some time
since one of the lucky citizens (Captain Vandergrift)
presented to one of the churches a bell, which was too
large for the belfry. A derrick had to be erected on
the premises, in which to hang the bell, and the whole
affair was so characteristic of the region that the
church was christened by some of the Oil creek boys
"The Church of the Holy Derrick," and is so recog-
nized.

THE FIRST STORE AND THE FOUNDER OF THE OIL CREEK ARISTOCRACY.

The first large store built in Oil City was erected
by a worthy gentleman by the name of John Hopewell,
who, with his amiable and accomplished daughters,
may be considered the actual founders of the Oil City
or petroleum aristocracy. They were the prime
movers in all the sociables, picnics, &c,, of the early
days of the place, and were regarded as the life and
soul of fashionable life. One of the daughters has
since married a gentleman in Meadville. John Hopewell is everywhere regarded as an upright, high toned
gentleman, and, with his family, are an ornament and
addition to respectable society anywhere. Dr. Bagg,
formerly from Michigan, and agent of the Michigan
Rock Oil Company; Mr. Lay, of Laytonia, and family,
formerly from Cincinnati; Messrs. Byles & Brown,
who opened the second store, were among the earliest
settlers who gave character to the place. Then came

several Michigan families, with the Michigan Company and with excellent taste and intelligence assisted in framing society from the crude elements by which they were surrounded. Since then several respected families from the eastward have settled in the place, and altogether society in Oil City is as refined as you will find it in any new country.

Among the most active business offices in Oil City is that of Fiske & Co., for the sale of oil lands, transfer of leases, interest in oil wells, &c. Mr. Fiske is a nephew of the late Judge Douglas ; the second member of the firm is Mr. H. S. Stevens, a well known and highly respected citizen of Cleveland, and the third is Mr. C. J. Fox, formerly United States Consul at Aspinwall. Mr. J. B. Chandler, of Lacon, Illinois, caught the oil fever at Oil City, and has sunk two wells in Livingston county, Illinois.

William A. Shreve, Esq., is also a prominent citizen of Oil City. He is President of the National Bank. William C. Tillson, Esq., an active resident is with Mr. Shreve. John H. Coleman, Esq., J. H. Winsor, Esq., (Franklin), Captain Vandergrift, Cornelius Curtis, Esq., George Barker, C. McKinley, Stephen H. Standart, of Franklin—a pioneer, very active —Joseph Bonefante, James Kincaid, young Crittenden, of Titusville—who got up the Crittenden Hall— and hundreds, perhaps thousands, of others, whose names it is impossible now to recall, and all hailing from various parts of Oil creek region, are among the lively spirits of the territory, and very successful in their business

PERSONALS

Among other gentlemen who have made money out of oil and speculation in lands in this region may be enumerated the following :

Mark A. Perry, resides in Utica, New York, formerly in Rouseville, made a pile of money out of the McClintock farm, part of which property has just been sold for $160,000. He is a gentleman of family, highly esteemed, enjoys life in a reasonable way, and is a whole souled representative of the petroleum aristocracy.

Hamilton McClintock, of McClintockville, a pioneer in the region, first gathered the oil in a rude way, and sold it for years; owns several steamers on the Alleghany, has a fine residence, a large family (eight or nine children), is robust and hearty, and as honest a looking man as you will see in a day's walk in Oil or any other creek.

Captain B. R. Alden, formerly on General Scott's staff, and wounded in the service of the United States, among the first to come here and enter into the oil business, bought an interest on the Buchanan and McClintock farms; realized a fortune and still retains a large portion of his property; is a high-toned, honorable gentleman, and lives in style in the city of New York.

George M. Mowbray, formerly of London, now of New York, the first agent sent out by New York parties— Schieffelin Brothers—to the oil region, is now in Titusville. Interested in wells, and puts np the best and most expensive works in these drillings. Has one of the best of families, and is liked for his good sense, liberality and public spirit.

James Parker, Titusville, formerly a merchant in New York city, bought property in Titusville. It became valuable, and he is now rich, hospitable, has great suavity of manners, a family, and is much esteemed.

J. W. Sherman, of Titusville, formerly of Saratoga, New York. Once a poor man, now worth a half a million; the original owner of the celebrated flowing

well known as the "Sherman well," which at one
time came near overflowing the whole region with
liquid bitumen. Is an unassuming gentleman, with a
pleasant, honest face and most agreeable manners.
Has a family. Occasionally spends some time in New
York. A very active business man.

Daniel Gregg, owner of the Gregg farm, below the
Miller farm, was poor, and hardly keeps pace with the
progressive spirit of the age. Farm not much developed. Asks four hundred thousand dollars in gold
for it : and is very shy when any one offers to purchase
it at his own price. Was at the Astor House last
spring, and proved himself to be a very domesticated
person, considering his means.

Robert Miller, of Miller's farm, four or five miles
below Titusville. Is a perfect specimen of a native,
a bachelor about forty, being too modest to speak to a
young woman ; lives with his mother ; hair brilliant
auburn ; honest as the longest day of the year, strictly
conscientious, and with a comical expression of visage
worthy of imitation by any delineator of the humorous
on the stage. Unprotected females can appear about
honest Bob Miller's drillings with perfect safety.
The Indian Rock Oil Company is located on this farm.

John Brown came to the pleasant village of Pleas-
antville, about five miles east of Titusville, from New
York city, about thirty years ago, and entered into the
general merchandise business. Mr. Brown prospered
well, and his sons, of the firm of Brown Brothers, now
carry on a most extensive business in banking, buying and selling real estate, and have just completed a
fine private residence for their mother. Samuel Q.
Brown, one of the firm, is interested largely in oil territory, and is an active member of a number of petroleum companies. All the brothers are active business
men, of lively dispositions ; very rich, young, not bad

looking, and, what will be more interesting to your young lady readers, bachelors. Unlike Bobbie Miller, however, it is not to be supposed that they are ever frightened at a pretty, modest face. Last November Mr. Brown bought the John Duncan farm, sixty-eight acres, above Titusville, on Pine creek, for $1,100. He has just disposed of it to Massachusetts parties for the sum of twenty thousand dollars ; reserving a quarter of the stock. It is a new region, just being developed. Some wells have been sunk.

C. J. Lloyd, Esq., is among the most energetic operators in the oil region. He is tireless in industrious works, and esteemed one of the most reliable men in the country. He is largely interested in oil territory, and does a heavy business in New York, Philadelphia, &c.

Mr. Funk, of Titusville, owner of the Funk farm, a very rich oil territory, the proprietor of Funkville, &c., is regarded as among the wealthiest young men in the whole region. He is administrator of the valuable estate of his father, Captain Funk, and is largely interested in oil territory in other parts of the country, particularly in a large tract in the State of Kentucky. His fortune may be set down at a million and a half.

F. Prentice, Esq., is also a very large and influential operator in oil lands. An active and reliable gentleman, good natured, and warm hearted and intelligent. His word is as good as " the bond of union"—perhaps better. Is very wealthy, and enjoys a very high reputation for integrity both in New York, where he transacts business, and at home among his neighbors.

George H. Bissell, Esq., of Franklin, Pa., formerly from New Orleans, realized a splendid fortune from early operations in the Oil creek region, and is entitled to much credit for his personal and untiring exertions in developing the oil territory.

[L. of C.

M'KINLEY OIL COMPANY

INCORPORATED UNDER THE LAWS OF THE STATE OF N. YORK.

Wells on Oil Creek, Pennsylvania.

CAPITAL,.....................25,000 SHARES.

PAR VALUE OF SHARES,   TEN DOLLARS EACH.

Morris Franklin, New York.   John H. Coleman, Oil City Pa.
James N. Lawton,   "   C. McKinley,   "   "
Sidney Cornell,   J. J. Vandergrift,   "   "
George Davis, New York.

President, MORRIS FRANKLIN.

Secretary, H. B. HENSON.

Treasurer, WALTER E. LAWTON.

Superintendents, McKINLEY BROS.

Office, 81 John Street, New York.

CLIFTON PETROLEUM COMPANY

INCORPORATED SEPTEMBER 1, 1864,

UNDER THE LAWS OF THE STATE OF NEW YORK.

Wells on Cherry Run and Cherry Tree Run, Penn'v'ia.

CAPITAL, 50,000 SHARES.
PAR VALUE OF SHARES, $10 EACH.

TRUSTEES.

James C. Wilson, New York,   John T. Daly, New York.
James N. Lawton,   "   John H. Coleman, Oil City, Pa.
Sidney Cornell,   "   David D. Hammond, "   "
George Davis, New York.

President, JAMES C. WILSON.

Secretary, H. B. HENSON.

Treasurer, WALTER E. LAWTON.

Superintendent, J. B. CHICHESTER.

Office, 81 John Street, New York.

New York and Liverpool Petroleum Co.

CAPITAL STOCK, $1,000,000.

100,000 SHARES AT $10 EACH.

Lands already yielding largely.

President, Hon. DANIEL S. DICKINSON.
Vice President, WILLIAM T. PHIPPS.
Secretary, ROBERT BASSETT.

Books are open for subscription at the office of the Company, 71 Broadway, Empire Buildings (Room 24), New York.

The lands of this company are situated in the heart of the oil regions, and include portions of those well-known localities, the McElheny farm, the two McClintock farms, and other proved and valuable working terory, including

Over Two Thousand Acres in Fee

Of the Best Oil Territory along Oil Creek and in West Virginia,

Now under Process of Successful Development, and

Oil is Largely and Regularly produced from several Wells upon them.

Address   **The New York and Liverpool Petroleum Co.,**
Box 5,368, Post-office, New York City.

M'KINLEY NO. 2 OIL COMPANY,

INCORPORATED OCT. 13, 1864,

UNDER THE LAWS OF THE STATE OF NEW YORK.

Wells on Oil Creek, Pennsylvania.

Capital, 25,000 Shares—Par value of Shares, $10 each

TRUSTEES.

J. E. Southworth (Pres't Atlantic Bank), New York.
E. S. Rich (Messrs. Rich & Sherman, Bankers), New York.
Rich'd L. Franklin (Yonkers & N. Y. Ins. Co.),   "
John Crompton, 35 Liberty street,   "
J. J. Vandergrift (First National Bank), Oil City, Pa.

President, J. E. SOUTHWORTH.
Treasurer and Secretary, R. W. R. Freeman.

OFFICE, ATLANTIC BANK, NEW YORK CITY

ALL ABOUT PETROLEUM.

A Weekly Journal

DEVOTED TO THE DEVELOPMENT OF THE PETROLEUM INTEREST.

The object of the publication of this Journal is to furnish all the information available on Petroleum, Petroleum Lands; and Petroleum Stocks, which is anywhere to be found.

It is got up in the very best style, and with a view of enabling the subscriber to have it bound; so that this Journal will form a complete history of the growth of the Petroleum Interest for future reference.

It is for sale by all first-class Newsdealers throughout the country. The trade is supplied by

THE AMERICAN NEWS Co., 121 Nassau-street, N. Y.

Subscriptions, at $5 per year, should be addressed to the publishers,

C. PFIRSHING & Co., 34 Liberty-street, New York.

Germania Petroleum Company,

Office, 33 Pine Street, New York.

Incorporated under the Laws of the State of New York, September 7, 1864.

Capital, 600,000 shares,

At the nominal par value of $5 00 each, issued in payment of property.

This Company has nine wells now producing, and fourteen wells partially down, some of which may become productive at any moment. Add to this the large amount of territory in possession, upon which may be sunk from one hundred to two hundred wells; and as it is the policy of this Company to push this development with great rapidity and vigor, it must be evident to stockholders that they may reasonably look for largely increasing dividends.

www.ingramcontent.com/pod-product-compliance
Lightning Source LLC
Chambersburg PA
CBHW021956190326
41519CB00009B/1289